Sources in the History of Mathematics and Physical Sciences

10

Sources in the History of Mathematics and Physical Sciences

Vol. 1: G.J. Toomer (Ed.)
Diocles on Burning Mirrors: The Arabic Translation of the Lost Greek Original, Edited, with English Translation and Commentary by G.J. Toomer

Vol. 2: A. Hermann, K.V. Meyenn, V.F. Weisskopf (Eds.)
Wolfgang Pauli: Scientific Correspondence I: 1919–1929

Vol. 3: J. Sesiano
Books IV to VII of Diophantus' *Arithmetica:* **In the Arabic Translation Attributed to Qustā ibn Lūqā**

Vol. 4: P.J. Federico
Descartes on Polyhedra: A Study of the *De Solidorum Elementis*

Vol. 5: O. Neugebauer
Astronomical Cuneiform Texts

Vol. 6: K. von Meyenn, A. Hermann, V.F. Weisskopf (Eds.)
Wolfgang Pauli: Scientific Correspondence II: 1930–1939

Vol. 7: J.P. Hogendijk
Ibn Al-Haytham's *Completion of the Conics*

Vol. 8: A. Jones
Pappus of Alexandria Book 7 of the *Collection*

Vol. 9: G.J. Toomer (Ed.)
Apollonius *Conics* **Books V to VII: The Arabic Translation of the Lost Greek Original in the Version of the Banū Mūsā,** Edited, with English Translation and Commentary by G.J. Toomer

Vol. 10: K. Andersen
Brook Taylor's Work on Linear Perspective

Kirsti Andersen

Brook Taylor's Work on Linear Perspective

A Study of Taylor's Role in the History of Perspective Geometry. Including Facsimiles of Taylor's Two Books on Perspective.

With 114 Illustrations

Springer Science+Business Media, LLC

Kirsti Andersen
History of Science Department
University of Aarhus
Ny-Munkegade
DK-8000 Aarhus C
Denmark

Brook Taylor's *Linear Perspective* and *New Principles of Linear Perspective* were printed for R. Knaplock at the Bishop's Head in St. Paul's Church-Yard in 1715 and 1719 respectively.

Printed on acid-free paper.

Typeset by Asco Trade Typesetting Ltd., Hong Kong.

9 8 7 6 5 4 3 2 1

ISBN 978-1-4612-6951-9 ISBN 978-1-4612-0935-5 (eBook)
DOI 10.1007/978-1-4612-0935-5

Brook Taylor
From *Principles of Linear Perspective* (ed. J. Jopling).

Preface

The aim of this publication, consisting of three "books", is to make Brook Taylor's two important works on perspective accessible, and to account for Taylor's influential role in the development of the mathematical theory of perspective and its applications.

Book One contains a presentation of Taylor's contributions to the theory of perspective and surveys their place in the history of this discipline from Guidobaldo del Monte to Johann Heinrich Lambert; it furthermore describes how significantly they guided the English generation of scientists and practitioners who worked on perspective after Taylor.

In Book Two, Taylor's *Linear Perspective* (1715) is reproduced to scale 110%. Book Three is a facsimile of his *New Principles* (1719), in which the figures have been reduced for technical reasons. Each of the last two books are organized so that the plates with figures are placed after the text, and they end with a section of notes where I have made cross references between Books Two and Three and references to Book One. Furthermore, I have added mathematical details to a few of Taylor's arguments and listed some misprints. In the margin to Taylor's text asterisks are placed to indicate my notes.

At the end of the publication is an index covering all three books.

Acknowledgments

Several have generously helped me in preparing this publication. Mette Dybdahl and Kate Larsen have put much work into eradicating linguistic mistakes, typing the manuscript, drawing the figures, and retouching the photographs of Taylor's books. Henk Bos, Jeremy Gray, Jesper Lützen, and Gerald J. Toomer have given very useful comments on the content and on the English of Book One. I am extremely grateful to all of them, and to my sons, Christian and Michael, for supporting this work.

Moreover, I am very thankful to the Danish National Library of Science and Medicine for kind permission to reproduce *Linear Perspective* and *New Principles* from the originals in its possession.

Finally, I want to thank the National Portrait Gallery, London, and the Danish National Library of Science and Medicine for permission to reproduce the photographs of Figures 1, 37, and 38.

Contents

Book One

Brook Taylor's Role in the History of Linear Perspective

Kirsti Andersen

Of all the Mathematical Sciences,
the study of Perspective is
perhaps the most entertaining.
MALTON[1]

1. Introduction

In 1715 Brook Taylor published his *Linear Perspective: or, a New Method of Representing Justly All Manner of Objects* (pp. 71–136). Responding to the criticism that the book was too concise Taylor revised it and in 1719 issued his *New Principles of Linear Perspective* (pp. 147–243). Although Taylor used the word new in the titles of both books he did not explain what he considered the novelty of his approach to perspective. The titles, however, had the effect that later in the century there came to exist an idea of a particularly Taylorian way of dealing with perspective. Thus the period from 1754 to 1803 witnessed the publication of at least seven English books on perspective with titles mentioning Dr. Brook Taylor's principles or method of linear perspective, and in the same period other English authors acknowledged that they had been inspired by Taylor. (More details are given in Section 9.) As late as the 1880s, in a textbook on perspective, the headmaster of the School of Art in Manchester, George O. Blacker, stated that Taylor was "the father of all modern Perspective" (Blacker, 1885–1888, Preface).

The interest in Taylor's work on perspective was particularly noticeable in England, but it also spread to the continent resulting in one French and two Italian editions of *New Principles* (Taylor, 1755, 1757, 1782). Furthermore, the idea of a special Taylorian approach to perspective was still alive in Italy in 1865 when Luigi Cremona wrote *I principii della prospettiva lineare secondo Taylor* containing new proofs of some of the fundamental theorems of *New Principles.*

When these evident proofs of Taylor's influence on perspective are compared with the fact that the theory of perspective was flourishing before he wrote on the subject, the question "What made Taylor's contributions so well

[1] Malton, 1779, p. ii.

known and acknowledged?" naturally arises. Guided by the wish to answer that question, I shall present Taylor's work on perspective in this introductory essay to the republication of his *Linear Perspective* and *New Principles*; furthermore, I shall relate it to the work of his contemporaries and finally investigate how the next generation of English writers applied his ideas.

It will turn out that Taylor enriched the theory in several respects, but that it is extremely difficult to distinguish a Taylorian theory of perspective. Moreover, it will appear that Blacker would have been right if he had characterized Taylor as the father of *English* perspective.

Taylor's contributions to perspective are mentioned in general descriptions of the history of perspective, such as the one given by Gino Loria in Moritz Cantor's *Geschichte der Mathematik* (Loria, 1908). A detailed examination of Taylor's work, however, is only found—as far as I am aware—in Phillip S. Jones's dissertation (1947). Unfortunately, this veritable goldmine of information on the history of perspective has not been published. Many of the factual statements in my exposition can also be found in Jones's dissertation; he and I, however, concentrate on different aspects of the development of the theory of perspective.

2. Taylor's Interest in Perspective

In his late twenties and early thirties Brook Taylor (1685–1731) had a very productive period. After being a fellow of the Royal Society for two years he was appointed its secretary in 1714, the year that he also graduated as Doctor of Law from Cambridge. He was a frequent contributor to the *Philosophical Transactions* on mathematical and physical topics throughout the years 1712–1724.[2] In 1715 he published, besides *Linear Perspective*, his most famous work, *Methodus incrementorum directa et inversa*. One of the concepts described in this book was later termed a Taylor series, and immortalized Taylor's name in the history of the calculus (Feigenbaum, 1985).

Taylor was also vividly interested in and practiced music and painting. According to his first biographer, his grandson William Young, Taylor was a skillful landscape painter:

> His [Taylor's] drawings and paintings preserved in our family require not those allowances for error or imperfection with which we scan the performances of even the superior *dilettanti*:—they will bear the test of scrutiny and criticism from artists themselves, and those of the first genius and professional abilities. [Taylor, 1793, p. 15]

[2] For more biographical information, see Taylor, 1793, Auchter, 1937, Jones, 1976, and Feigenbaum, 1985 and 1986.

Figure 1. Taylor showing the manuscript of *Linear Perspective*, the painter is supposed to be Joseph Goupy.

So far, I have not been able to trace any of Taylor's paintings; one of them, however, has been reproduced in the portrait of Taylor shown in Figure 1.

In his biography, Young stressed that he himself did not understand mathematics—he even seemed to have held a particular view about people who do so, for he characterized Pierre Remond de Monmort as "the lively, yet learned mathematician" (Taylor, 1793, p. 9). Hence Young supplied no information of the kind we would like to have about the connection between Taylor's interest in painting and the mathematical theory of perspective. But it seems reasonable to assume that Taylor—like so many before and after him—was taught the rules of perspective without any scientific explanation. Being a mathematician, Taylor might have wanted to understand the *rationale* behind these rules, and thereby was led to the study of the theory of perspective. Besides, he had an interest in synthetic geometry which in 1711 he expressed in the following manner:

> ... it is my opinion that the prevailing humour of treating Geometry so much in an Algebraical way has prevented many noble discoveries, that might otherwise have been made, by following the Methods of the Ancient geometricians.[3]

Taylor's first presentation of the result of his work on perspective, *Linear Perspective*, does not reveal any reflections on the existing literature; but in *New Principles* he was more informative:

> Considering how few, and how simple the Principles are, upon which the whole Art of PERSPECTIVE depends, and withal how useful, nay how absolutely necessary this Art is to all sorts of Designing; I have often wonder'd, that it has still been left in so low a degree of Perfection, as it is found to be, in the Books that have been hitherto wrote upon it. [p. 149 = Taylor, 1719, p. iii]

Taylor's complaint was not new; in almost all prefaces of seventeenth-century books on perspective we find similar statements; and it was repeated—as we shall see in Section 9—after Taylor's books had appeared. Unfortunately, Taylor did not tell which books he had consulted and found unsatisfactory. But he claimed that:

> it seems that those, who have hitherto treated of this Subject, have been more conversant in the Practice of Designing, than in the Principles of Geometry .... [p. 150 = Taylor, 1719, p. iv]

Thus he did not disclose any familiarity with books written by able mathematicians. In 1732 a catalogue was issued for the sale of the books left on the deaths of Joseph Hall and Brook Taylor. Among them are several—most likely stemming from Taylor's collection—in which perspective is presented quite adequately as a part of geometry (Appendix, p. 67). So Taylor presumably read what some of his colleagues had written on the subject, if not before 1715, then later. One of the best books preceding Taylor's own two books, W.J. 's Gravesande's *Essai de perspective*, is not included in the catalogue, nevertheless, as I shall argue in Section 4, it seems beyond doubt that Taylor was inspired by this work. 's Gravesande's approach to perspective was very mathematical and that was precisely what Taylor considered to be the proper one, even for nonmathematicians. Thus he suggested the following instruction of a painter:

> I would first have him learn the most common Effections of Practical Geometry, and the first Elements of Plain Geometry, and common Arithmetic. When he is sufficiently perfect in these, I would have him learn *Perspective*. [p. 158 = Taylor, 1719, p. xii]

Not only should the painter be acquainted with perspective, he should learn

---

[3] Taylor to Mr. Newcome, 24 November, 1711, unpublished letter in the Library of St. John's College, Cambridge, quoted from Feigenbaum, 1986, p. 53.

it well:

> Nothing ought to be more familiar to a *Painter* than *Perspective*; for it is the only thing that can make the Judgment correct, and will help the Fancy to invent with ten times the ease that it could do without it. [p. 159 = Taylor, 1719, p. xiii]

The request for a perspectively correct representation was so basic for Taylor that he made the following comparison:

> A Figure in a Picture, which is not drawn according to the Rules of *Perspective*, does not represent what is intended, but something else. So that it seems to me, that a Picture which is faulty in this particular, is as blameable, or more so, than any Composition in Writing, which is faulty in point of Orthography, or Grammar. [p. 155 = Taylor, 1719, p. ix]

With the attitude that perspective, and a geometrical understanding of it, was a *sine qua non* for painting, Taylor developed his theory of perspective; and he did not give his reader the impression that there exists a royal road to his theory:

> The Reader, who understands nothing of the Elements of Geometry, can hardly hope to be much of the better for this Book .... [p. 154 = Taylor, 1719, p. viii]

In Section 9 we shall see that in the second half of the eighteenth century Taylor's ideas were taken up by English practitioners. They found, however, that a less abstract way than Taylor's, of transmitting geometrical knowledge, was needed.

3. The Basic Concepts of Taylor's Method

The aim of perspective has often been confusingly described in the literature; in particular, it seems to have been difficult to distinguish sharply between a reproduction of the visual impression of a figure and a reproduction of the figure that gives the eye the same visual experience as the figure itself (cf. Andersen, 1987_1, p. 81). For Taylor there was no doubt about the function of a perspective picture: it

> ought so to appear to the Spectator, that he should not be able to distinguish what is there represented, from the real original Objects .... [p. 161 = Taylor, 1719, p. 1]

The means through which Taylor wanted to achieve this was—as we saw in the previous section—geometry. This discipline had been a tool for painting since 1435, when Leon Battista Alberti presented the first mathematical model for making perspective drawings in a plane. Alberti defined the perspective image of a plane figure as the intersection of the picture plane and the pyramid, called the visual pyramid, which has the figure as base and the eye point of an observer as vertex. Taylor took over this definition replacing the visual

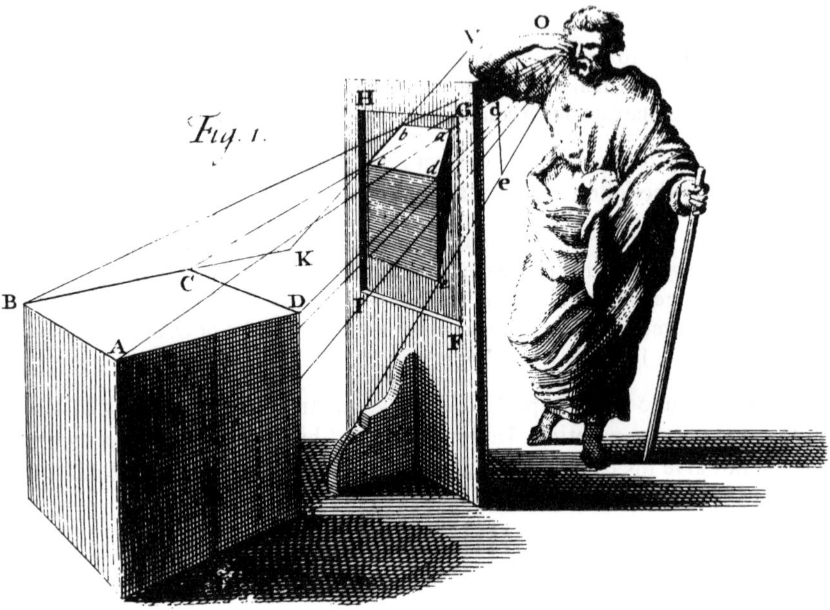

Figure 2. Taylor's illustration of the mathematical model for perspective drawings. The face *ABCD* of the cube is depicted in *abcd*, where the point *a* is determined as the point of intersection between the visual ray *OA* and the picture plane, and the other points are found similarly. *New Principles*, Figure 1 (p. 231).

pyramid with the analogous concept of an "Optic Cone" (p. 163). Thus, expressed in modern terms, for Taylor a perspective image of a figure was the image obtained by performing a central projection, from an eye point, of the figure upon a picture plane. Taylor named the image the *Representation*, but in *New Principles* he also employed the term *Projection*. Taylor's illustration of how a cube is projected upon the plane *HGFI* from the eye point *O* is reproduced in Figure 2.

Taylor did not touch upon the fact that he—like all other writers on perspective—applied a model in which the complicated process of seeing with two movable eyes is approximated by one in which all light rays are assumed to converge in one fixed eye point. For him the important thing was to solve, as elegantly and concisely as possible, the geometrical problem of constructing the projections of various figures.

Taylor was of the opinion that it was "absolutely necessary to consider this Subject [perspective] entirely anew" (p. 150). In reforming the subject he introduced several concepts that were either completely new or had been little used. One in particular is interesting, namely the concept of a vanishing line, which is a generalization of the concept of a vanishing point. The latter had occurred—under various names—in mathematical treatises on perspective from the very beginning of the seventeenth century. The vanishing point of a line, not parallel to the picture plane, is the point where a line through the eye point parallel to the considered line meets the picture plane. (This definition implies that parallel lines have the same vanishing point.) The vanishing point of a given line has the fundamental property that it lies on that line into which the given line is depicted (the vanishing point is actually the image of the point at infinity of the given line).

Taylor used—and seems to have been the first to do so—the term vanishing point and gave two reasons for it. The first, which seems rather artificial, was that a line that passes through its vanishing point will be depicted in just that one point (Figure 3), "and may be said to vanish" (p. 175). The second and more intuitive was (Figure 3) that the further away on a line a line segment of a given length is from the picture plane, the smaller it will be depicted and the nearer its image will be to the vanishing point, "and when it comes into this Point, its Magnitude vanishes because the Original Object is at an infinite Distance" (*ibid.*).

Taylor was—as we shall see in the next section—very much aware of the fact that the concept of a vanishing point was extremely useful in dealing with perspective images of lines. This realization may have inspired him to generalize the concept, and assign a *vanishing line* to a given plane which is not parallel to the picture plane; thus he defined the vanishing line as the intersection of the picture plane and the plane through the eye point parallel to the given plane. Taylor did not invent the concept of a vanishing line; it occurs implicitly in the early mathematical literature on perspective, namely as a line containing all the vanishing points belonging to different sets of parallel lines that are all parallel to a given plane. Guidobaldo del Monte (1600, p. 46) emphasized particularly that the vanishing points of horizontal lines lie in a line—later called the *horizon*—whereas François Aguilon (1613, p. 651) noticed the collinearity of vanishing points of lines situated in a nonhorizontal plane. Defined explicitly as lines belonging to planes, vanishing lines are to be found in Charles Bourgoing's book (Bourgoing, 1661, p. 5). Nevertheless, Taylor was the first to let vanishing lines—different from the horizon—play an important role in the theory of perspective and was presumably also the first to term them vanishing lines.

In his books Taylor gave no reason for introducing the concept of a vanishing line—nor did he emphasize the novelty of using it. But in an account of *Linear Perspective* he explained that:

> our Author has rejected the term of *Horizontal Line*, because it confines the Mind too much to the particular considerations of the Horizontal Plane: but he considers all Planes alike. [Taylor, 1715_2, p. 301]

From a mathematical point of view it is sensible to treat all planes in the same way. But it is an abstraction that was of no help for the practitioners who were used to dealing with a horizontal plane of reference. Taylor did not completely give up the horizontal plane, as can be seen in Figure 4, but he retained his general approach by assuming that the picture plane is not vertical.

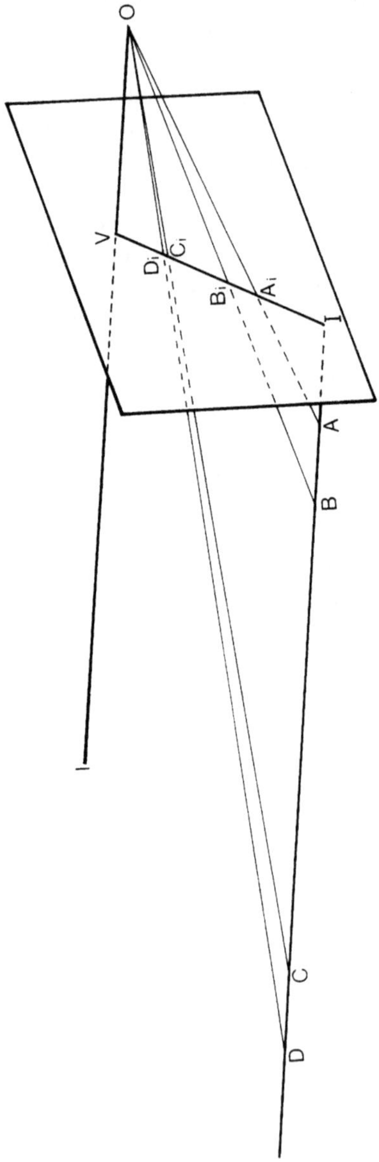

Figure 3. The parallel lines *l* and *m* have the vanishing point *V*, *l* passes through the eye point *O*, and *m* intersects the picture plane at *I*. The image of *l* is just the one point *V*, whereas *m* is depicted into the line *IV*. The line segments *AB* and *CD* are equal, but their images A_iB_i and C_iD_i are unequal, and the further away we imagine *CD* from the picture the smaller its image will be, and the nearer to *V*.

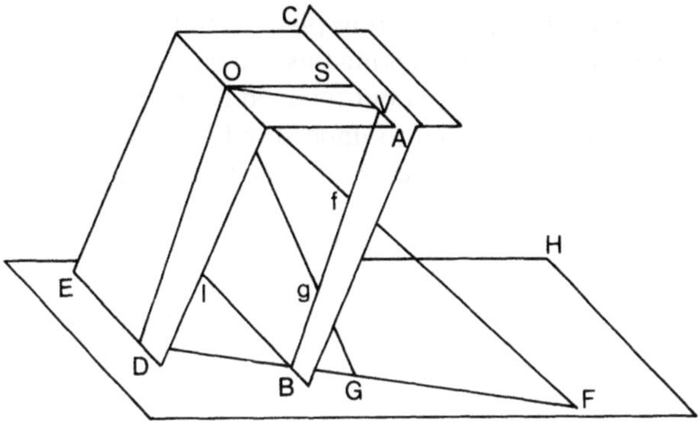

Figure 4. A corrected version of Taylor's diagram illustrating the basic concepts of the theory of perspective. *O* is the eye point, *FGH* is a horizontal original plane, and *ABC* is an oblique picture plane. The line *OV* is parallel to *GF* and thus *V* is the vanishing point of the line *GF*. *New Principles*, Figure 3 (p. 232).

Taylor used Figure 4 to illustrate all his concepts. I shall present only those necessary for a further understanding of this survey of his work, and one which is particularly remarkable. The point *O* is the eye point—called the *Point of Sight* by Taylor, although he presumably took the *O* from *oculus*—the plane *ABC* is the picture plane, called the *Picture*. The orthogonal projection of *O* onto the picture Taylor termed the *Center of the Picture* (not marked in the figure); for the distance between *O* and *ABC* he took over the usual name: the *Distance of the Picture*, often just called the *distance*. This definition implies that the center is the vanishing point of lines orthogonal to the picture plane. In a traditional situation, where the picture is assumed to be vertical and where especially images of horizontal lines are studied, this vanishing point is much used and is sometimes called the principal vanishing point. In Taylor's general approach the center plays a less important role as a vanishing point.

By an *Original Line* Taylor understood a line—like *GF*—whose projection is to be found; its intersection with the picture plane he simply called the *Intersection*. Analogically, he described the concepts of an *Original Plane* and its *Intersection*. In Figure 4 the intersection of the plane *FGH* is the line *BI* and *FGH*'s vanishing line is *AC*—it follows from the definition of the latter that these two lines are parallel. The point, *S*, where the perpendicular from *O* to *AC* intersects *AC*, is called the *Center of the Vanishing Line*, while the length *OS* is called the *Distance of the Vanishing Line*.

In Figure 4 we furthermore notice the plane *OED* which passes through the eye point and is parallel to the picture plane; Taylor termed this the *Directing Plane*—some later writers termed it the vanishing plane. Taylor introduced several other concepts connected to the directing plane, but actually made very little use of these and the plane itself—they only occur in a few alternative perspective constructions in *Linear Perspective* (pp. 83, 96) and *New Principles* (pp. 177, 187, 190). Thus Taylor's motive for setting up a special terminology connected to the directing plane is not clear to me. One guess might be that he had thought of making this plane a central tool in his theory, and that he later gave up this idea but retained his nomenclature.

As to the mathematics needed to understand his proofs Taylor assumed, as mentioned earlier, that his readers had some knowledge of plane and solid geometry, and he used Euclid's *Elements* as a reference work. In *New Principles*, however, he decided to let some of the incidence relations occur as axioms (pp. 170–171, for more details, see Bkouche, 1990, pp. 23–26).

In quoting Taylor I have used *New Principles*; all the described concepts can, however, also be found in *Linear Perspective*. I shall continue to base my investigations of Taylor's theory of perspective mainly on *New Principles* because Taylor in general expresssed himself most clearly in this book, and because it was through this that most of his successors became acquainted with his theory. It should be remarked, however, that although *New Principles* is the best to consult for most subjects, *Linear Perspective* has the advantage of being more coherent. Thus it is easier to distinguish the idea behind Taylor's choice and arrangement of the theorems, problems, and examples in *Linear Perspective* than in *New Principles*. Furthermore, there are—as pointed out by Jones—a few mathematically interesting considerations that can only be found in *Linear Perspective* (Jones, 1951).

4. Taylor's Inheritance

In this section I shall deal with some fundamental parts of Taylor's theory that originate from earlier writers. The starting point will be the theorem Taylor formulated in the following way:

> The Projection of a straight Line not parallel to the Picture, passes thro' both its Intersection and Vanishing Point. [p. 174 = Taylor, 1719, p. 14]

In Figure 5 the theorem is illustrated with my symbols (in general I use the subscript *i* for image): The point O is the eye point, π the picture plane, l a line that is not parallel to π, I_l the intersection of l, and V_l the vanishing point of l. The theorem then states that the image, l_i, of l is determined by $I_l V_l$. Having proved this theorem Taylor adds in *New Principles*:

> This theorem being the principal Foundation of all Practice of *Perspective*, the Reader would do well to make it very familiar to him. [p. 174 = Taylor, 1719, p. 14]

Taylor's observation is extremely relevant, the theorem is indeed very central in the theory of perspective (cf. Andersen, 1984)—and I shall term it *the main theorem of perspective*. Since the beginning of the seventeenth century, mathematicians working on perspective had applied this theorem intensively, but usually without emphasizing its importance. Before Taylor, the English math-

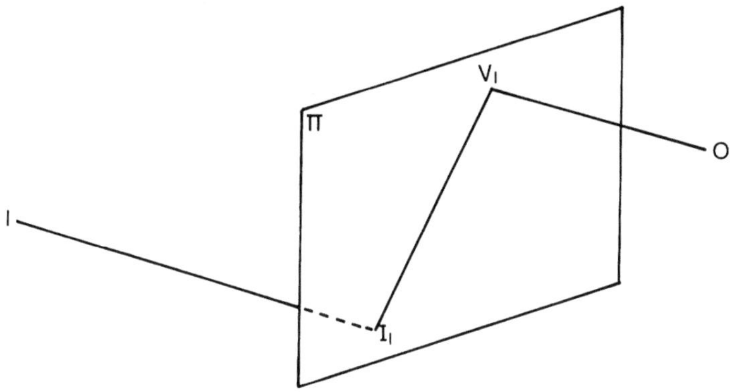

Figure 5. An exemplification of the main theorem.

ematician Humphry Ditton had stressed that the result, that all parallel lines—which are not parallel to the picture plane—have the same vanishing point, is "the main and great Proposition" (Ditton, 1712, p. 45). But Taylor realized that the combination of this insight with the fact that an intersection serves as its own image was the real fundamental result in the theory of perspective.

The main theorem was particularly instrumental for the derivation of constructions of perspective images of plane figures and for the proofs of their correctness. Already Guidobaldo del Monte, who was the first to publish a form of the main theorem, developed 23 different constructions of the perspective images of a point by means of it (Guidobaldo, 1600, pp. 61–105).

Among all the many constructions one especially was preferred by the practitioners, namely the so-called distance point construction. This came to exist long before the main theorem was established but had not been proved to be correct. The main theorem made it an esay task to do so, for the construction was based on the idea of considering a point as a point of intersection of two lines of which one is at right angles and the other at an angle of 45° to the picture plane, and then using the intersections and vanishing points of these lines. The latter are the principal vanishing point and a point called a distance point because its distance to the principal vanishing point is equal to the *distance* of the picture.

Taylor did not show any interest in this popular construction, but took up less-known ones. His first construction is simple and nice—in it he used the term *seat* for the orthogonal projection of a point onto the picture plane:

PROBLEM I

Having given the Center and the Distance of the Picture, to find the Projection of a Point, whose Seat on the Picture, with its Distance from it, are given. [p. 180 = Taylor, 1719, p. 20]

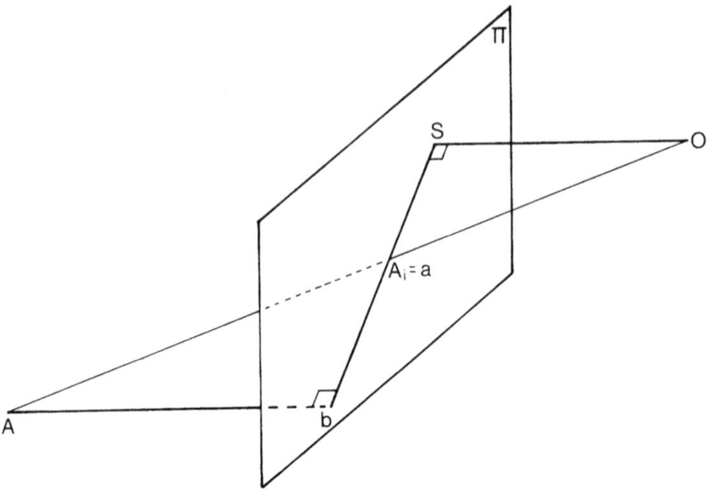

Figure 6. Illustration to Taylor's Problem I.

In Figure 6 I have illustrated the three-dimensional configuration, the point O is again the eye point and π the picture plane; S is the center of the picture—i.e. the orthogonal projection of O onto π—A is the given point and b its seat. It is supposed that besides the points S and b the distances OS and bA are given; the image A_i of A has to be determined. Since the vanishing point of the line Ab is S and its intersection is b, it follows from the main theorem that the line segment Ab is depicted on a part of Sb, and in particular that A_i lies on Sb. Using similar triangles we get further that

$$SA_i : bA_i = OS : bA. \tag{4.1}$$

From this Taylor concluded that A_i can be determined by any construction that performs a division of the line segment Sb in the ratio $OS : bA$. Such constructions have the advantage of being very easy to carry out directly in the picture plane; Figure 7 shows one of the solutions suggested by Taylor (note that the point O is no longer the eye point but a point in the picture plane).

A perspective construction, based on a division of the line segment between the intersection and the vanishing point of a line, had been used before Taylor,[4] but he made it a more fundamental tool. *Inter alia* he used it as a foundation for an extremely elegant construction which, like the previous one, served to determine the images of points. Lines like OA in Figure 7 play an important role in the construction, and since Taylor called such lines visual

[4] We find it, for instance, in Tacquet, 1669, Ozanam, 1693, Lamy, 1701, and 's Gravesande, 1711.

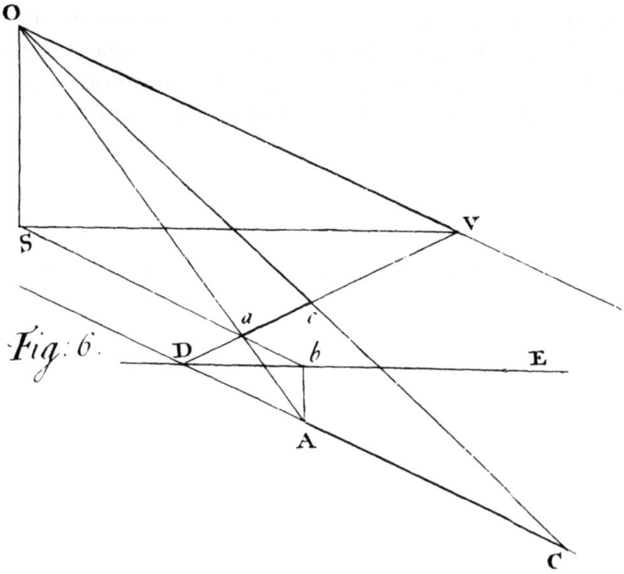

Figure 7. Taylor's construction of Problem I; his explanation reads: "Let S be the Center of the Picture and b the given Seat of the Original Point. Draw at pleasure SO equal to the Distance of the Picture, and parallel to it draw bA equal to the Distance of the Original Point from its Seat. Draw Sb and AO meeting in a, which will be the Projection sought". *New Principles*, page 20 and Figure 6 (pp. 180 and 232).

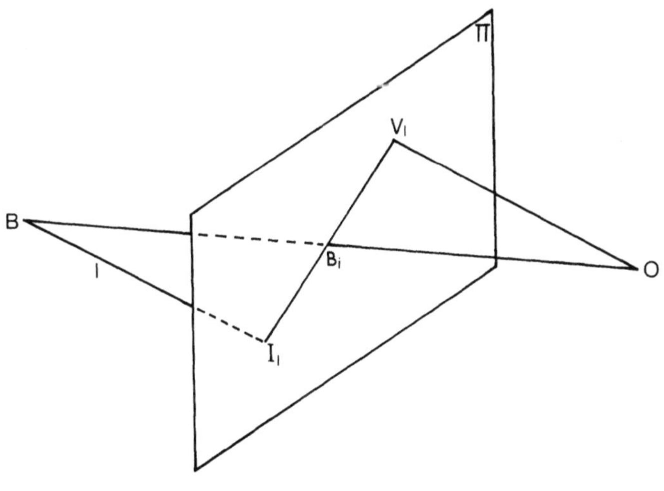

Figure 8. A generalization of Taylor's Problem I.

rays (I shall return to this expression), I term the construction the *visual ray construction*. Taylor often applied the construction in *New Principles*. However, rather than emphasizing its importance by letting it occur as a separate problem, he made it a corollary of Problem II (p. 182). I shall spend a few more words than Taylor on explaining the construction. First, we observe that for the division procedure of Problem I the assumption that *OS* and *bA* are perpendicular to the picture plane is not essential. Let (Figure 8) *B* be any given point on a line, *l*, whose intersection, I_l, and vanishing point, V_l, are given; since V_l is the vanishing point of BI_l, the lines OV_l and BI_l are parallel,

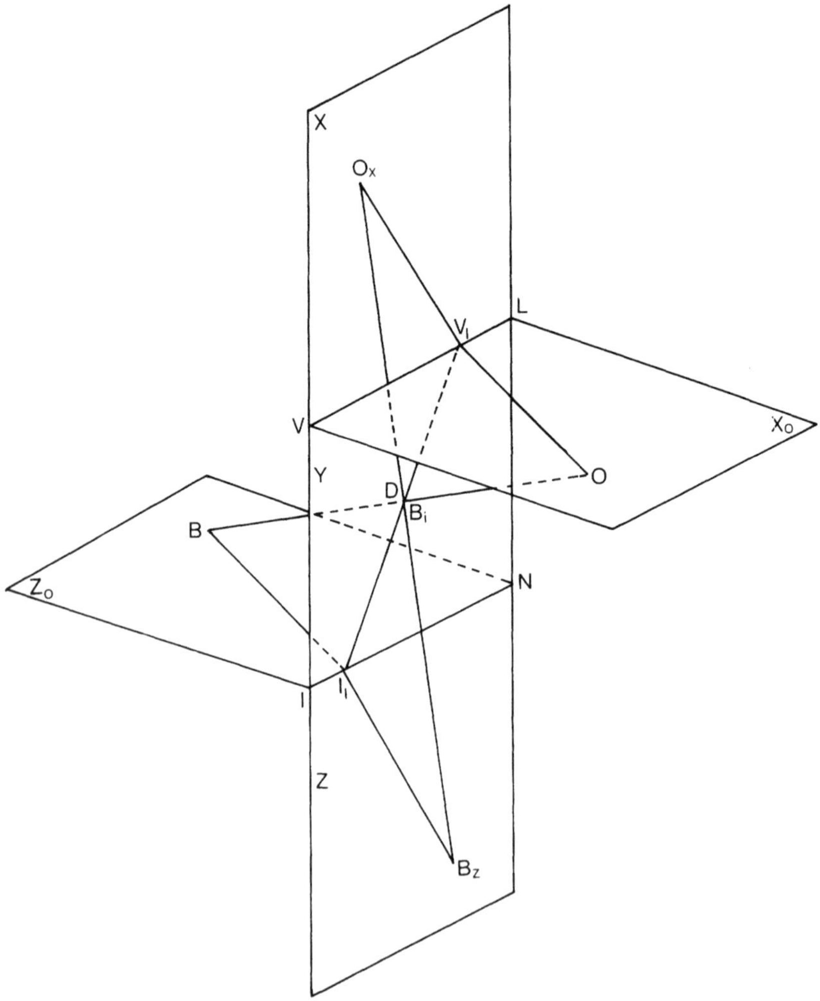

Figure 9. Illustration to the visual ray construction.

hence the image, B_i, of B is determined by the relation

$$V_l B_i : I_l B_i = OV_l : BI_l. \tag{4.2}$$

The visual ray construction is based on the idea of rotating the original points as well as the eye point into the picture plane. This is illustrated in Figure 9, where O is the eye point, B is an original point that lies in a plane Z_0 whose intersection, IN, and vanishing line, VL, are given. (Inspired by Jones, I term intersections of planes IN and vanishing lines VL (Jones, 1947).) Let X be the part of the picture plane which is above VL, Y the part between VL and IN, and Z the part below the latter line. Furthermore, let l be a line passing through B and having the intersection I_l and the vanishing point V_l;

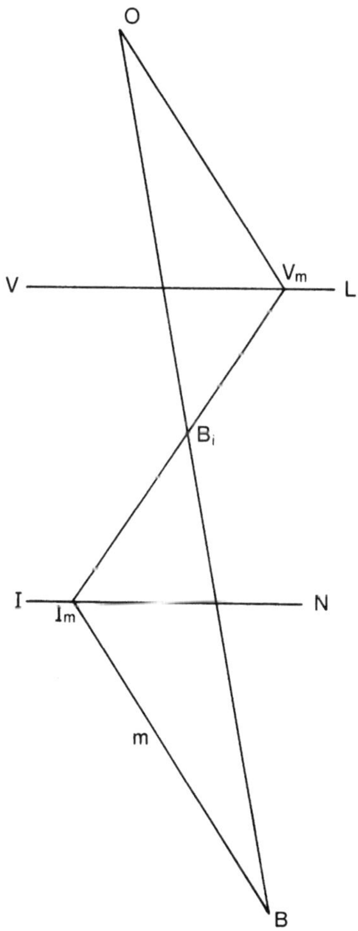

Figure 10. The visual ray construction.

and let finally X_0 be the plane through O and VL—which means that X_0 is parallel to Z_0. We now rotate X_0 around VL onto X and similarly Z_0 around IN onto Z; the point O will then fall on O_x and B on B_z.

The aim of the visual ray construction is to determine the image B_i of B by using only the points O_x, B_z, I_l, and V_l which are all in the picture plane. The idea of the construction is the fundamental observation that the points O_x, B_i, and B_z are collinear. To realize this Taylor hinted at a proof where the parallelism of OV_l and BI_l is used to deduce that O_xV_l and B_zI_l are parallel, and that therefore the point of intersection, D, between O_xB_z and I_lV_l will divide I_lV_l in the ratio $O_xV_l : B_zI_l$. Since $OV_l = O_xV_l$ and $BI_l = B_zI_l$, the point D divides V_lI_l in the same ratio as the point B_i (cf. (4.2)), which means that the two points coincide, and hence B_i lies on O_xB_z. (Another way of reaching this conclusion is to observe that O_x is the vanishing point of the line BB_z whose intersection is B_z; therefore O_xB_z is the image of BB_z and contains in particular the point B_i.)

As already indicated, Taylor used the expression the *visual ray* for the line O_xB_z. This terms is slightly misleading for it may give the wrong impression that O_xB_z is obtained by rotating OB—around B_i—into the picture plane. Nevertheless, in naming the construction I have respected Taylor's choice.

Though it is a little complicated to explain, the visual ray construction itself

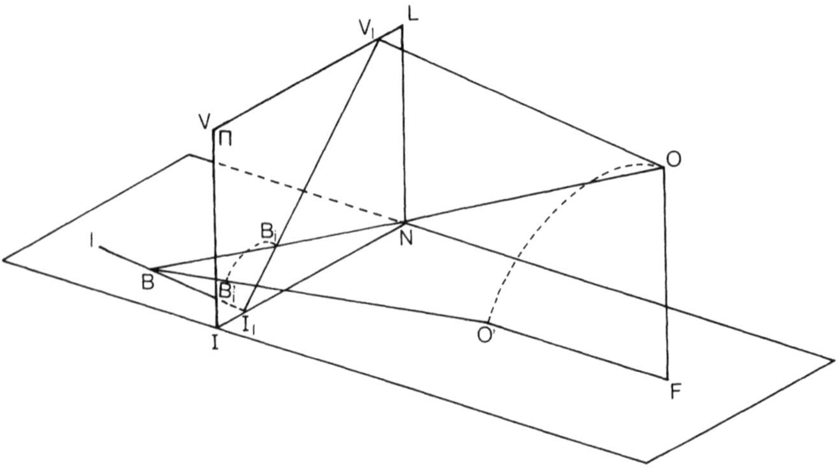

Figure 11. An illustration of Stevin's investigation of rotations (the notation is mine). π is the picture plane; O is the eye point; F is its orthogonal projection into the ground plane. The line l also lies in this plane with Intersection I_l and vanishing point V_l; B is a point on l and B_i its image. Stevin imagined that O and π are rotated simultaneously around F and IN, respectively, so that OF stays parallel to π. He then proved that the rotated B_i will continuously be the image in the rotated picture plane of the point B seen from the rotated O. How Stevin used this result to derive a perspective construction can be seen in Figure 12.

is very simple to perform when the following is given (Figure 10): the position of a point B with respect to the intersection, *IN*, of a plane; the vanishing line, *VL*, of this plane; the center of the vanishing line (i.e. the orthogonal projection of the eye point on the vanishing line); and the distance between the eye point and the vanishing line. To find the perspective image B_i of *B*, we determine B_z and O_x from the data—but now call them *B* and *O*—and draw the line *OB*, then through *B* we draw any line *m* which is different from *OB* and intersects *IN*, find its intersection I_m, and finally we draw a line through *O* parallel to *m* and find its point of intersection, V_m, with *VL*. The point where $V_m I_m$ and *BO* meet will then be the required B_i.

The procedure of rotating a ground plane—like Z_0 in Figure 9—into the picture plane was already used in the sixteenth century and gradually became a standard procedure in perspective constructions. The unusual step in Taylor's construction is that he also rotated the eye point into the picture plane. The actual idea of using invariance of the perspective image of a point

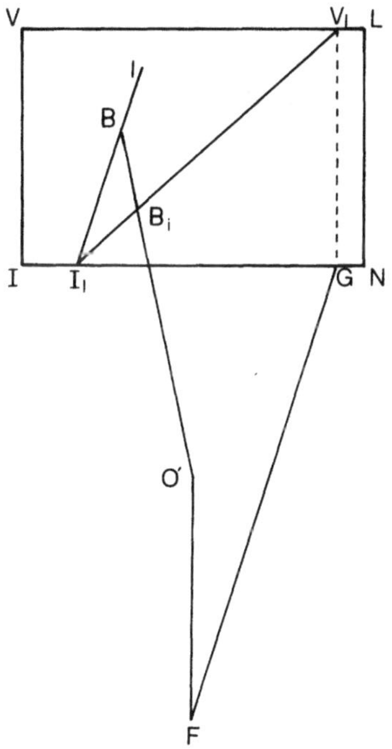

Figure 12. This diagram illustrates the situation where the rotations described in connection with Figure 11 have brought *O* and π to the ground plane (which in this diagram has been made to coincide with the plane of the paper). Stevin's result mentioned above combined with a continuity argument implies that B_i lies on $O'B$; according to the main theorem it also lies on $I_l V_l$, and therefore B_i is the point of intersection between $O'B$ and $I_l V_l$. I_l can be found immediately as the point of intersection between *IN* and *l*; V_l is found by drawing *FG* parallel to *l* (*FG* is then parallel to OV_l, Figure 11) and draw GV_l at right angles to *IN*.

under rotations—where the eye point is rotated—originates from Simon Stevin who used it to develop a simple and elegant construction in *Van de verschaeuwing* (Stevin, 1605, pp. 29–31). Thus by rotating the eye point and the picture plane into the ground plane Stevin achieved the construction illustrated in Figures 11 and 12. Stevin's and Taylor's procedures are rather different. Stevin's rotations result in a configuration where the ground plane and the picture cover each other and the eye point is situated below the vanishing line *VL* of the ground plane, whereas the result of Taylor's operation is that the ground plane lies below the line of intersection *IN* of the ground plane and the picture, and that the eye is rotated into an auxiliary eye point situated above *VL*.

An auxiliary eye point like Taylor's is uncommon in the literature preceding his work, but Taylor was not the first to introduce this point, nor was he the first to use the "visual ray" between the auxiliary eye point and the original point rotated into the picture plane (Jones, 1947, pp. 95–96). This line was especially much used by 's Gravesande in his *Essai de perspective* (1711) for deducing various constructions of the perspective image of a point. One of these constructions is identical to Taylor's visual ray construction; even the use of the symbols *X*, *Y*, and *Z* for the three parts of the picture plane is shared

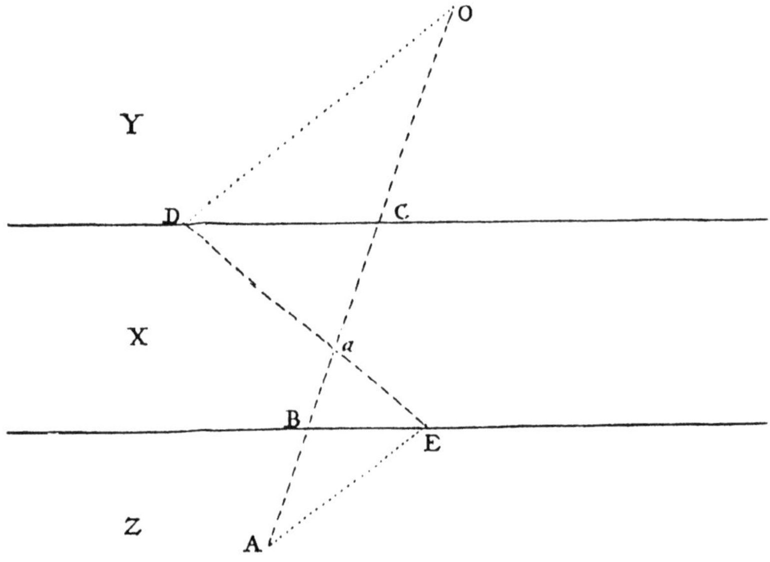

Figure 13. 's Gravesande's visual ray construction. *O* is the eye point rotated into the picture plane, *DC* is the horizon, *X* is the picture, *BE* is the ground line, and *Z* the ground plane turned into the picture plane. The construction is the same as Taylor's, thus the image *a* of the point *A* is found in the following way. Through *A* draw an arbitrary line *AE* (different from *AO*), draw *DO* parallel to it, and then find *a* as the point of intersection between *OA* and *DE*. *Essai de perspective*, Figure 8.

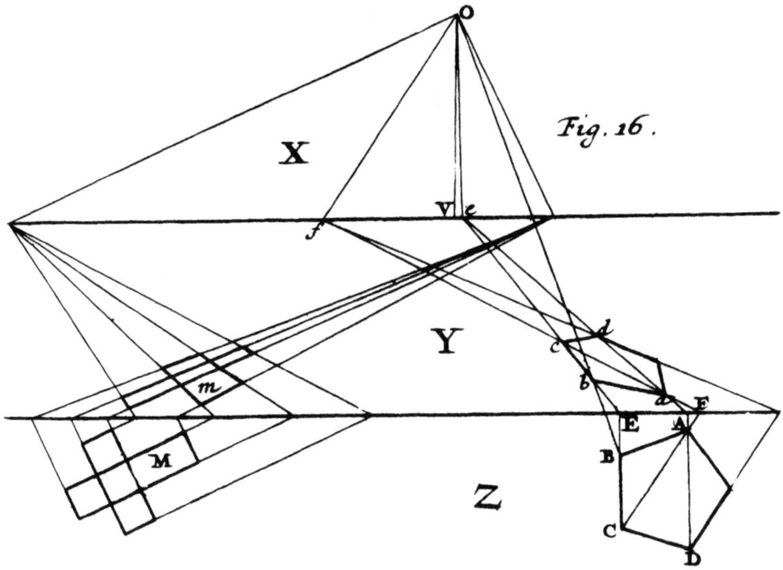

Figure 14. One of Taylor's compositions where he, as 's Gravesande did in the previous figure, has divided the plane in three parts and denoted them *X*, *Y*, and *Z*. The right part of the diagram illustrates how Taylor used the "visual ray" *OB* to construct the image *b* of the point *B*. There is a drawing mistake in the diagram as *Oe* is supposed to be parallel to *BE*. *Linear Perspective*, Figure 16 (p. 121).

by the two mathematicians, although they chose different orders (Figures 13 and 14). Another similarity between Taylor's and 's Gravesande's presentations is remarkable. 's Gravesande stressed— which was not common—that a procedure like the one used in the visual ray construction implies that the constructions be performed on the side of the *tableau* which faces the original object rather than the one facing the eye. This has, he said, " the same effect as if one after having made a drawing looked at it from behind."[5] Taylor, likewise, remarked that the figure is "seen on the back-side" (p. 96) or "seen on the Reverse, as Objects appear in a Looking Glass" (p. 186).

As also pointed out by Jones a few more of Taylor's constructions resemble some given by 's Gravesande in *Essai de Perspective* (Jones, 1947, pp. 115–118; another example of likeness is shown in Figures 15 and 16). The similarities are too striking to be accidental, thus it seems beyond doubt that Taylor was inspired by 's Gravesande to some extent when he wrote his books. Among 's Gravesande's several constructions Taylor recognized the visual ray construction as very elegant, and contrary to 's Gravesande he decided to

[5] 's Gravesande, 1711, p. 28: "le même effet que si après avoir fait un dessein, on le regardoit par derrière."

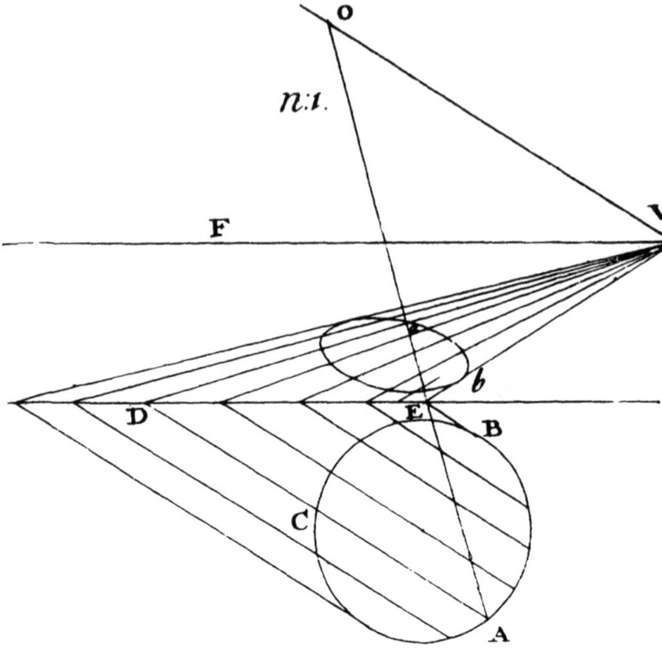

Figure 15. Taylor's construction of the problem of determining the images of some points on the circle *ABC*. In the diagram we can see how Taylor applied the "visual ray" construction to find the image, *a*, of the point *A*, thus *a* is the point of intersection between the "visual ray" *OA* and the line *VD* which, according to the main theorem, is the image of the line *AD*. Taylor did not draw the other "visual rays" used for throwing other points of the circle into perspective. He has, however, illustrated how he economized by using chords parallel to *CA*, their images all passing through *V*. *New Principles*, Figure 13 (p. 235).

make it a rather fundamental tool for throwing plane figures into perspective (Figure 15). Taylor also showed his independence by not copying 's Gravesande's proof of the visual ray construction; 's Gravesande used the main theorem twice to deduce (Figure 9) that B_i lies on O_xB_z, whereas Taylor—as we saw—established this from the fact that O_xB_z divides V_lI_l in the right ratio.

The examples dealt with in this section illustrate a general trend in the way Taylor used his predecessors' work on perspective. He picked out the most fruitful ideas and made them basic in his own theory. Moreover, his approach was more mathematical in the sense that he deduced more general results than his predecessors. Many of his results are admirable from a mathematical point of view. For his readers, however, they may also have been rather confusing because he gave no guidance in seeing what these results should be used for. The succeeding three sections contain more examples of Taylor's ability to sort out the mathematically interesting problems in the theory of perspective—and his shortcoming in making his aims clear.

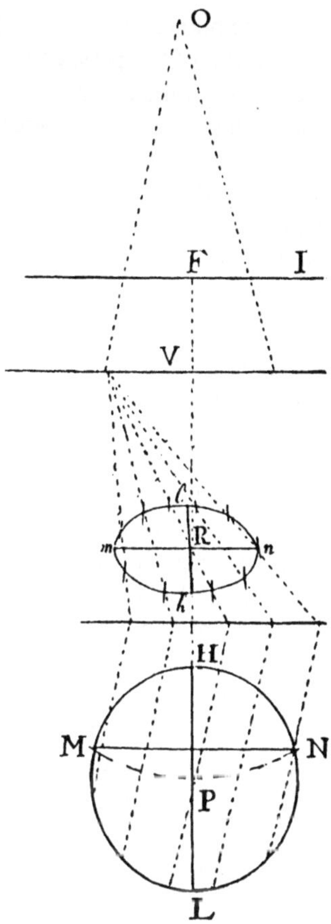

Figure 16. An illustration of how, before Taylor, 's Gravesande used parallel chords in the process of putting a circle into perspective (cf. Figure 15)—before him Guidobaldo had also used this method (Guidobaldo, 1600, p. 217). The diagram is a part of Figure 16 in *Essai de Perspective* (reproduced from the 1774 edition; the original one from 1711 is similar).

5. Towards a Perspective Geometry

In this and the following section I shall deal with Taylor's contributions to perspective geometry. To evaluate these I shall, however, first briefly outline the history of this discipline.

Among the numerous perspective constructions invented after Alberti had presented the first in 1435, those which were used most required a plan, that is, the orthogonal projection of an original object—in some scale—into a horizontal ground plane. The plan was often—as mentioned in the previous section—rotated into the picture plane, and in that plane its points served as auxiliary points in the construction of the perspective images of points in the original object. This procedure was taken under review by Johann Heinrich Lambert in the 1750s. He found that is was too cumbersome as it meant double work: first a plan of the object had to be produced, and thereafter the plan had to be thrown into perspective (Lambert, 1759_1, pp. 2–3). The procedure furthermore has the disadvantage that it is necessary to adjust the scale of the plan to the one used in the picture plane. This problem, however, was considered a minor one and is generally ignored in the literature on perspective (cf. Andersen, 1987_2).

Lambert decided on a new approach where he wanted to "free" perspective constructions from the plan. His studies resulted in a very fascinating book, which appeared in 1759 in a German edition as *Die freye Perspektive* and in a French edition as *La perspective affranchie de l'embaras du plan géométral.*[6] To avoid the plan Lambert used the method of laying out angles and line segments directly in the picture plane. Furthermore, he performed some other constructions—which are basic in the Euclidean plane—directly in the picture plane, an example being the constructon of a line through a given point parallel to a given line. Combining these ways of "freeing" perspective, Lambert reached his great and important achievement which applied to three-dimensional space and which he himself termed *perspective geometry* (Lambert, 1759_1, p. 12). I shall take over this name; moreover, I shall call constructions that can be performed directly in the picture plane *direct* or *free* constructions, and I shall occasionally refer to the picture plane as the perspective plane.

Lambert was the first to set himself the task of always operating directly in the picture plane and to conceive this as having its own geometry. Most of his ideas, however, had been used separately by previous authors. The very first book on the theory of perspective, Guidobaldo's *Perspectivae* from 1600, already contained the germs to a perspective geometry. Thus Guidobaldo showed how a given angle can be constructed directly in the picture plane, he gave a direct solution to the above-mentioned problem of determining a line through a given point parallel to a given line, and he dealt with the problem of dividing a line segment in the picture plane in a given ratio (Guidobaldo, 1600, pp. 112–114, 233–234). A quarter of a century later the French engineer Jacques Aleaume had the idea of making direct constructions for perspective images of horizontal line segments and angles and for vertical line segments (Aleaume, 1643).[7] The French seventeenth century literature contains several other examples of direct constructions, the most noticeable being in the works

[6] On Lambert's contribution to perspective, see Lambert, 1943, 1981, and Laurent, 1987.

[7] Although it is unclear which of the various results in Migon's edition of Aleaume's work originate from the author and which from the editor, I find it likely that the original approach has remained unchanged and hence assume that the idea of performing direct constructions was Aleaume's.

by Charles Bourgoing (1661) and Grégoire Huret (1670) (cf. also Jones, 1947, pp. 85–87). Girard Desargues's perspective method, which implies a construction of a coordinate system in the perspective plane, can also be considered as an example of working without a plan (Desargues, 1636).

Despite the Frenchmen's interesting free constructions there is no doubt that the most important figure in the prehistory of perspective geometry is Brook Taylor. It is not his attitude that makes him important, because he did not show any sign of a conscious attempt to "free" perspective—although S.N. Michel later claimed so,[8] nor did Taylor distinguish between free and other constructions. Taylor is important because his theoretical considerations contain all the material necesssary for creating a perspective geometry.

In placing Taylor in the development of a perspective geometry it would be extremely relevant to know to what extent he was inspired by his predecessors and what role his own work played for Lambert. Unfortunately, these points are not easy to clear up. Taylor's procedures for direct constructions are—as far as I am aware—his own, hence it is difficult to tell whether he was stimulated by others to treat this kind of problem or whether he took them up independently. Similarly, it seems impossible to answer the question whether Taylor not only anticipated Lambert but also inspired him. It is certain that at some point in his life Lambert became acquainted with Taylor's *New Principles* and had a copy of the French edition from 1757 in his book collection (Lambert, 1943, p. 48), but whether he knew the content of *New Principles* before he wrote his own book—which appeared in 1759—cannot be decided. In a chapter on the history of perspective—which he added to the second edition of *Die Freye Perspective* (1774)—Lambert mentioned Taylor's work, but not exactly favorably:

> Taylor treats the theory [of perspective] very generally, because from the beginning he assumes that the picture is oblique. Besides that he uses mainly new and superfluously many concepts which—although they give him more theorems—make the theory unnecessarily complicated.[9]

Notwithstanding this reservation the possibility exists that Lambert had noticed the interesting approach of making direct constructions in Taylor's "complicated" theory before he planned his own book—which according to

---

[8] Thus Michel claimed:

> "C'est-là tout l'art de cette Perspective linéaire inventée par le célèbre Docteur Brook Taylor, Anglois. Elle a cet avantage sur les autres Méthodes qu'elle donne la projection perspective ... sans qu'on soit obligé ... d'en faire aucuns plans ni élévations géométrales ...." [Michel, 1771, p. 6]

[9] Translation from Lambert (1774, p. 29):

> "Taylor handelt die Theorie sehr allgemein ab, weil er die Tafel gleich anfangs als schiefliegend annimmt. Über dies hat er meistens neue und überflussig viele Benennungen, die ihm zwar mehrere Lehrsätze geben, dabei die Theorie ohne Notwendighkeit weitläufiger machen".

the notes in his *Monatsbuch* happened in September 1758 (Lambert, 1916, p. 21). However, it is just as likely that he first saw Taylor's work, which is not mentioned in the *Monatsbuch*, after he had created perspective geometry.

In turning to a description of Taylor's work on direct constructions, I shall start with the situation where a plane—which I call the *plane of reference*—is given and the perspective images of objects in this plane are sought. The constructions of these images can—like constructions in the Euclidean plane—be built up stepwise, and some basic constructions can be recognized. The most fundamental "semi-metric" construction is the following (Bricard, 1924, §24):

(1) To divide a given line segment in a given ratio.

Among the metric constructions the following two are particularly fundamental:

(2) To cut off a line segment of a given length on a given line from a given point.
(3) Through a given point on a given line to draw a line making a given angle with the given line.

Although Taylor never revealed his criteria for selecting problems his work betrays that he attempted to treat the most basic ones. Thus it is interesting to notice that his investigations led him close to solving the three above-mentioned problems in the perspective plane; or to be more precise, that he solved the first and third completely and the second in a slightly restricted form.

Before showing how Taylor solved these problems for the perspective case I shall—inspired by Lambert—introduce the concept of perspective equality to overcome the inconvenience that a central projection does not in general preserve equality between magnitudes. I say that two angles in the perspective plane are perspectively equal when they are images of equal angles in the plane of reference, and if the original angles are v degrees I say that the images perspectively are v degrees. Similarly, when a line segment in the plane of reference is divided in a given ratio I say that its image perspectively is divided in this ratio. Taylor did not introduce these concepts explicitly, but they underlie his vocabulary.

To be as general as possible, Taylor did not fix the plane of reference but formulated his problems so that they apply to a set of parallel planes of reference. This does not matter for the concepts just defined, because they are invariant for a change from one plane of reference to another parallel to it. Taylor's generality, however, excludes the possibility of assigning a perspective length to a line segment in the perspective plane. Hence he could not treat the exact perspective equivalent of the second problem but had to assume that the perspective length of one line segment on the line was given (Problem IV, p. 27). The problem of dividing a line segment perspectively Taylor formulated in the following way:

PROBLEM III

Having given the Projection of a Line, and its Vanishing Point; to find the Projection of the Point that divides the Original Line in any given Proportion. [p. 183 = Taylor, 1719, p. 23]

While Taylor often allowed lines to be indefinite, for instance, in his applications of the main theorem, he also used the term *line* in the classical sense of a line segment. The latter meaning is found, among many places, in Problem III and in the following perspective version of the second problem:

PROBLEM IV

Having given the Projection of a Line, and its Vanishing Point; from a given Point in that Projection, to cut off a Segment, that shall be the Projection of a given Part of the Original of the Projection given. [p. 184 = Taylor, 1719, p. 24]

Finally, he transformed the third problem to the perspective plane in the following manner:

PROBLEM XI

Having given the Vanishing Line of a Plane, its Center and Distance, and the Projection of a Line in that Plane; to find the Projection of another Line in that Plane, making a given Angle with the former. [p. 192 = Taylor, 1719, p. 32]

To solve Problem III Taylor performed a very simple construction (Figure 17): Given are the line segment *AB* in the picture plane and the vanishing point *V* on the line *AB*; moreover, there is given a ratio which I shall denote by $r:s$. From the point *V* Taylor laid down an arbitrary line segment *OV*, at an arbitrary distance he drew a line *ab* parallel to *OV*, and found the points, *a* and *b*, where it meets *OA* and *OB*. He then divided *ab* in the given ratio,

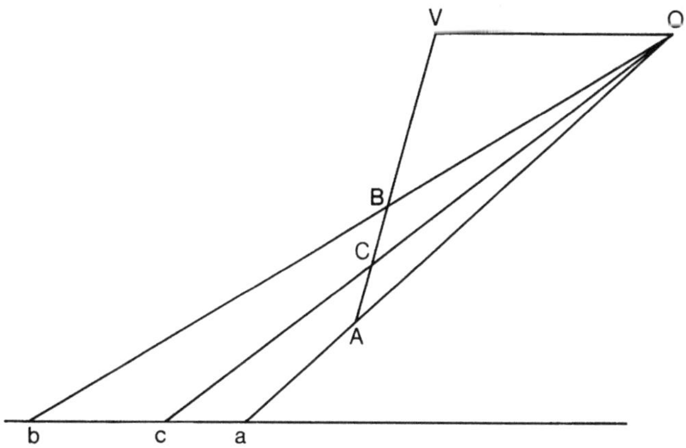

Figure 17. Taylor's construction of Problem III (cf. Figure 7, p. 233).

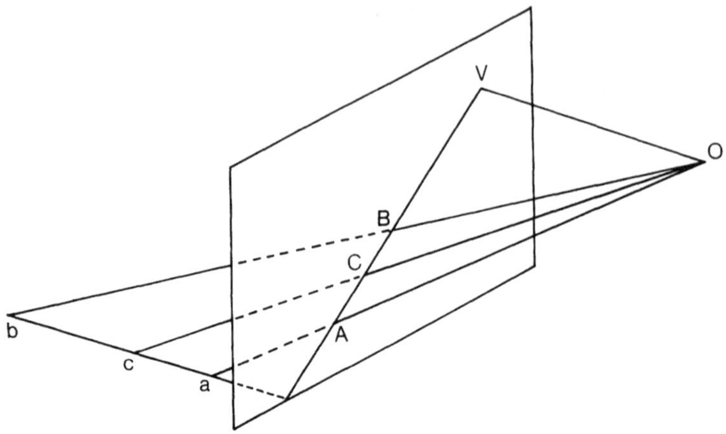

Figure 18. An illustration to Taylor's Problem III.

i.e. constructed c so that

$$ac : bc = r : s,$$

and maintained that the point of intersection, C, of AB and Oc is the required point. As was his habit Taylor did not spend too many words on the demonstration (in Figure 18 I have illustrated the three-dimensional configuration):

> OV being parallel to ba, ba may be consider'd as the Original Line and OV as its Parallel, and consequently O as the Point of Sight, and aO, bO, cO, as Visual Rays projecting the Points A, B, C. [p. 183 = Taylor, 1719, p. 23]

Taylor trusted that his readers understood that since the entire argument only concerns points in one plane, the construction can be performed directly in the picture plane. Furthermore, he assumed that the readers would realize that the position of C on AB is independent of the choices of O and the line ab, provided that OV and ab are parallel. To the "Mathematical Reader" Taylor remarked that he "will easily find" (p. 183) that

$$(AC \cdot BV) : (BC \cdot AV) = ac : bc.\text{[10]} \qquad (5.1)$$

This relation confirms the above-mentioned independence.

The magnitude $(AC \cdot BV) : (BC \cdot AV)$ was later termed the cross ratio of $ABCV$, and is an important concept in projective geometry. Commenting upon the result (5.1) Jones has observed that it "amounts to a use of the

[10] The relation (5.1) can be established in the following way (Figure 19). Through C we draw a line parallel to VO meeting Ob in F and Oa in E. From the set of similar triangles ACE, AVO and BVO, BCF we get

$$AC : AV = EC : VO \qquad (5.2)$$

and

$$BV : BC = VO : CF. \qquad (5.3)$$

Furthermore, since FE is parallel to ba we have that

invariance of the cross ratio of four points under a projection in which one point goes to infinity" (Jones, 1947, p. 98). Stretching Taylor's argument a shade, one may even say that his solution of Problem III implies that the cross ratio of four points is invariant under any central projection.[11]

Taylor solved Problem IV similarly to Problem III, but challenged his

---

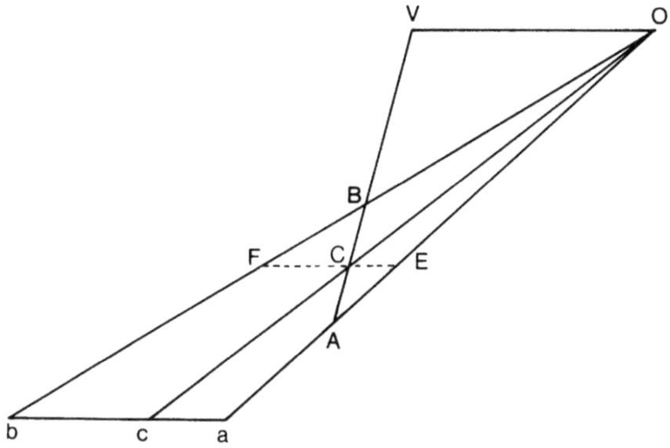

Figure 19

$$EC : CF = ac : bc. \tag{5.4}$$

By multiplying (5.2) and (5.3) and combining the result with (5.4) we obtain (5.1).

[11] Given that the four collinear points A, B, C, D are depicted by a central projection from a point O in the four collinear points A', B', C', D' (Figure 20), we want to show that

$$(AC \cdot BD) : (BC \cdot AD) = (A'C' \cdot B'D') : (B'C' \cdot A'D'). \tag{5.5}$$

We consider the two lines AB and $A'B'$ as two lines in a picture plane and assume that they have vanishing points D and D', respectively. Let ab be a line parallel to OD intersecting OA in a, OB in b, and OC in c. From Taylor's Problem III it follows that

$$(AC \cdot BD) : (BC \cdot AD) = ac : bc,$$

$$(A'C' \cdot B'D') : (B'C' \cdot A'D') = ac : bc,$$

and these relations imply the required result (5.5).

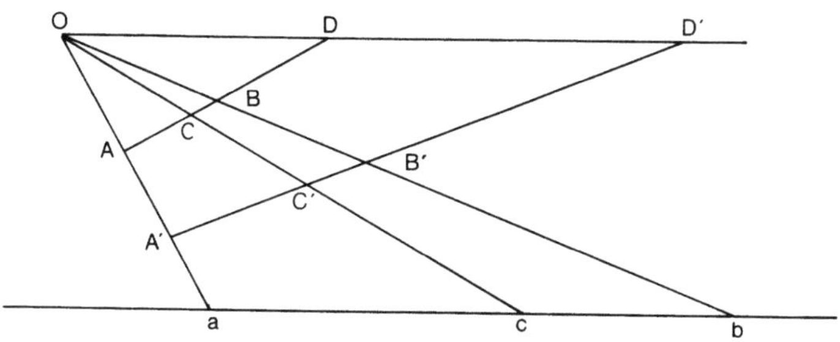

Figure 20

readers by establishing the solutions of Problem IV through a proof (pp. 184–185) that is different from and more complicated than that used for Problem III.

In dealing with problems concerning free constructions it would be natural to suppose that the positions of the eye point, the picture plane, and the original object were known. Taylor's approach, however, was that of a mathematician who does not assume more than necessary. It has alredy been mentioned, that he did not distinguish between parallel planes of reference; in formulating Problems III and IV he was, however, even more general and gave solutions which apply for any combination of eye point and original line when the latter is parallel to the line between the eye and the given vanishing point, and for any distance between the eye and the picture plane.

Taylor kept to his general approach while treating the topic of Problem XI where a line has to be constructed so that it perspectively makes a given angle with a line already constructed in the perspective plane. However, since an angle—i.e. a two-dimensional magnitude—is involved he had to make more assumptions than in Problems III and IV; and he assumed that a vanishing line, its center, and distance are given (cf. p. 11). The vanishing line given in Problem XI belongs to a plane which is determined by two original lines making the given angle, let us denote this plane by γ. Let, furthermore, η be the plane normal to the given vanishing line through its center; Taylor's construction in Problem XI then applies for all situations where the eye point lies on a circle in η that has the same center as the vanishing line, and as radius its distance and where γ is parallel to the plane determined by the eye point and the given vanishing line. In other words, if we choose γ first we can choose it as any plane whose intersection is parallel to the given vanishing line; γ and the distance will then determine the eye. On the other hand, if we choose the eye first we can choose it as any point on the circle described above, the eye and the given vanishing line will then determine the direction of γ (which means there is still a choice to be made among parallel planes).

Rather than paraphrasing Taylor's solution of Problem XI (pp. 192–193) and his proof that the solution is correct, I shall explain the underlying ideas (Figure 21). Let O be the eye point, VL the given vanishing line, AB the given perspective line segment, and let the given angle be denoted by v. Let us imagine that the problem is solved, which means that the angle BAC perspectively is equal to v, i.e. it is the projection of an angle baC which equals v and lies in a plane γ that has the vanishing line VL. Let, moreover, IN be the intersection of γ, F the vanishing point of ab, and G that of aC. Since ab is parallel to OF, and aC to OG, it follows that angle FOG equals v.

Reversing the argument we conclude that the line AC can be constructed in the following way: we determine the point of intersection, F, of AB and VL, make the angle FOG equal to v and join A and G; the line AG is then the required line. By rotating the eye point into the picture plane—as he did in the visual ray construction—Taylor performed the construction just described in the picture plane (Figure 22).

Already in Aleaume's and Huret's books on perspective we find the prob-

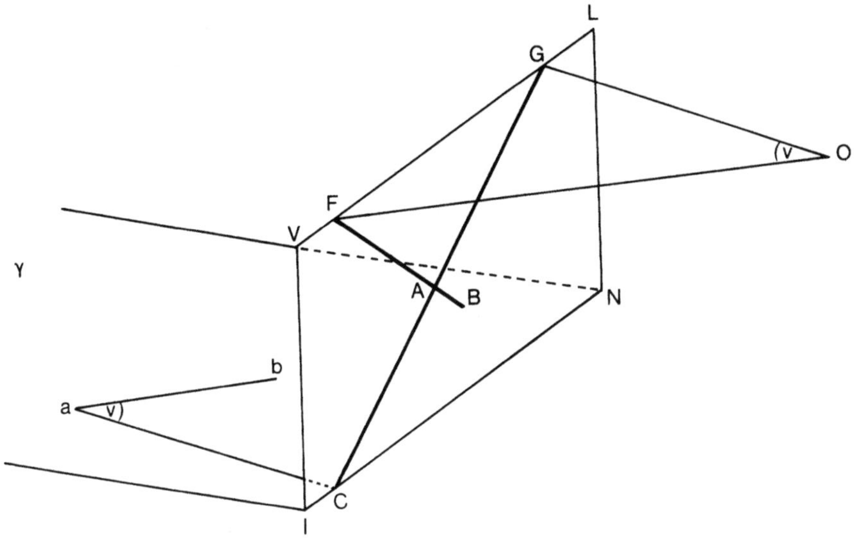

Figure 21. A diagram to Taylor's Problem XI.

lem of constructing angles directly in the perspective plane, and an extremely convenient tool for solving it—namely a scale on the horizon for measuring angles (Aleaume, 1643, pp. 73–75; Huret, 1670, p. 116). Later Lambert made this scale very instrumental in his theory of perspective, terming it *Winkelmesser* in German and a *transporteur perspectif* in French; when he first introduced it he thought it was his own invention. After his *Freye*

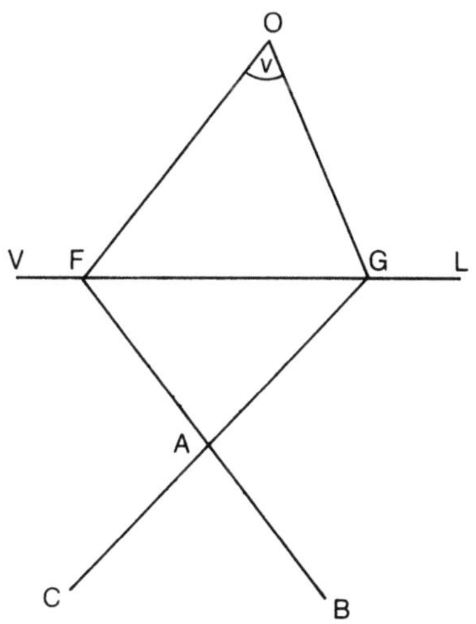

Figure 22. Taylor's construction of Problem XI (cf. Figure 10, p. 233).

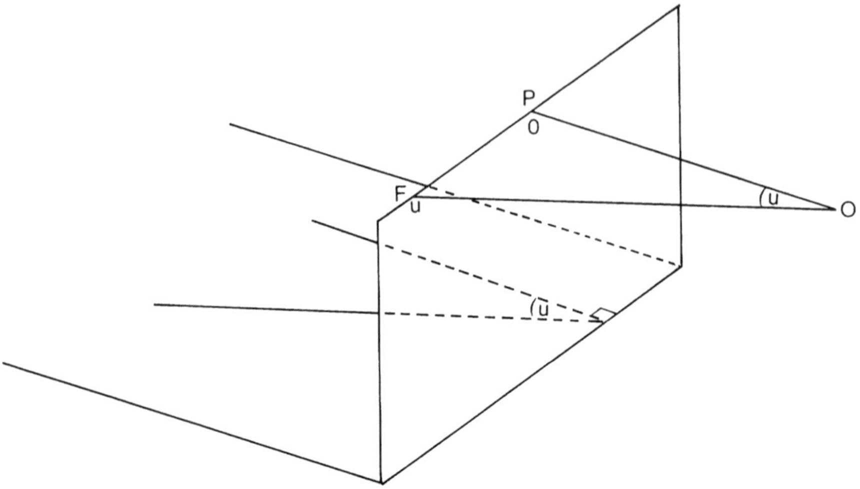

Figure 23. The construction of an angle scale.

Perspektive (1759) had appeared he realized that it also occurred in N.L. La Caille's *Leçons élémentaires d'optique* (1756, p. 145), but he remained unaware of the fact that he been preceded more than one hundred years earlier. Taylor was not among those who anticipated Lambert—but almost, because he had all the relevant observations for constructing an angle scale.

To be more specific on the point of how close Taylor was to the idea of a such a scale I shall first describe it. Let us for the sake of simplicity assume that the picture plane is vertical (Figure 23), that the eye point, O, is fixed, and that we have a horizontal plane of reference. To a point F on the horizon we assign the number u when F is the vanishing point of horizontal lines making the angle u (measured in degrees) with the normal to the picture plane. In particular, the orthogonal projection, P, of the eye point O is marked with a zero. For marking the other points on the horizon we can use the fact that

$$PF = OP \tan u. \tag{5.6}$$

In dealing with vanishing points Taylor several times invoked trigonometry and noticed the result corresponding to (5.6)—for instance, in *Linear Perspective* (p. 86) he observed what amounts to $PF = OP \cot(90 - u)$. Thus if Taylor had assigned numbers to points on the vanishing line for a given plane, his solution of Problem XI would have shown him the following (Figure 23): When the vanishing point F has been found, then G is determined by one of the requirements (depending on the size of v) that either the sum or the difference between the numbers at F and G must be v.[12] It was precisely this

[12] Let the vanishing point F of the given line in the picture plane have the mark u (Figure 24). We want to construct the angle BAC perspectively equal to v (where it is an implicit requirement that v is less than $90° + u$). Let D be the point of intersection of the line AB and the ground line.

solution that some of Taylor's predecessors, and later Lambert, used for the problem (Lambert, 1759_1, p. 14).

Reviewing the part of Taylor's work dealt with in this section we notice that he provided the theory for a method which comes close to free constructions of the images of figures in a given plane. As soon as the perspective image of one line segment of a given rectilinear figure, situated in a given plane, is known, the image of the figure can be completed by a direct construction based on the results of Problems IV and XI. Taylor explained this procedure in Example II in *Linear Perspective* (p. 97) and in Problem XIII of *New Principles* (p. 194). Moreover, in *New Principles* he applied this method to some examples concerning the perspective images of regular polygons (pp. 189, 195).

Linear Perspective, furthermore, contains three examples of direct constructions related to the circle. In the first two Taylor showed how one can construct as many points as one pleases on a perspective circle, if either its center and one of its radii or three of its points are given. The last example deals with a direct construction of tangents to a circle from a given point (p. 99). Undoubtedly, Taylor was aware that the perspective image of a circle is a conic section, but he did not—unlike several of his contemporaries—take up a study of these projections. In this matter he showed himself a practical man, claiming that it is preferable that the images of "Curve-lined Figures are described by finding several Points, and then joyning them neatly by Hand" (p. 98). He stressed that "this may conveniently be done by putting the

---

Since the angle at D in the triangle ACD is perspectively equal to $90° - u$, the angle at C is perspectively equal to $90° + u - v$. We distinguish between two cases:

(i) $v \geq u$. In this case, $90° + u - v = 90° - (v - u)$ is an acute angle and the original of AC makes the angle $v - u$ with the normal to the picture plane, but it is counted in the opposite direction to the angle u. Hence we conclude that the vanishing point of AC is the point G marked $v - u$ which lies on the opposite side of P from F.

(ii) $v \leq u$. In this case, $90° + (u - v)$ is an obtuse angle and the original of AC makes the angle $u - v$ with the normal to the picture plane, this time counted in the same direction as angle u. Hence the vanishing point G of AC is marked $u - v$ and lies on the same side of P as F.

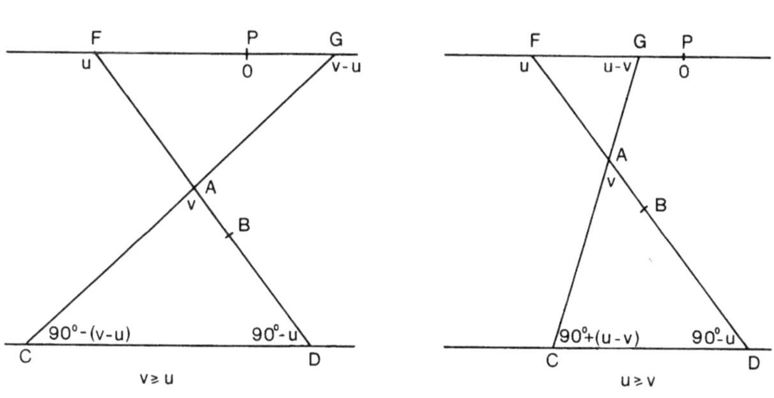

Figure 24

Geometrical Descriptions of Curves into Perspective" (*ibid.*)—i.e. by transforming the properties of the curves directly to the perspective plane. This idea is completely in accordance with the idea of a perspective geometry.

6. Three-Dimensional Perspective Problems

After the examination of Taylor's contributions to a two-dimensional perspective geometry it is natural to turn to the question of how he proceeded with three-dimensional figures. Traditionally, the perspective image of a three-dimensional figure was found in two steps; in the first the image of a plan of a figure was constructed, and in the second the images of points above the plan were found by using an elevation of the figure. The application of this procedure seems to have been in Taylor's mind when he showed—in *Linear Perspective*—how the plans and elevations of the five Platonic solids are found (pp. 100–103). However, Taylor did not restrict himself to this procedure but advanced in *Linear Perspective* a new method which he also described in *New Principles*; his idea was to generalize the direct constructions so that they could be applied to three-dimensional objects. Before illustrating Taylor's method I shall outline his basic results concerning three-dimensional direct constructions. By and large they correspond to solve the following problems in the Euclidean space:

(1) Determine the direction of the normals to a given plane (Taylor, 1719, Problem XIV, p. 197),

and vice versa:

(2) Determine the direction of the planes normal to a given line (Taylor, 1719, Problem XV, p. 199).

Moreover,

(3) Given a plane and a line that is not perpendicular to the plane, determine the normal plane to the given plane which contains the given line (Taylor, 1719, Problem XVI, p. 200).
(4) Given a plane and a line in it. Determine the plane through the given line that makes a given angle with the given plane (Taylor, 1719, Problem XVII, p. 201).

Taylor's perspective version of the first problem was the following:

PROBLEM XIV

Having given the Center and the Distance of the Picture, and the Vanishing Line of a Plane; to find the Vanishing Point of Lines perpendicular to that Plane. [p. 197 = Taylor, 1719, p. 37]

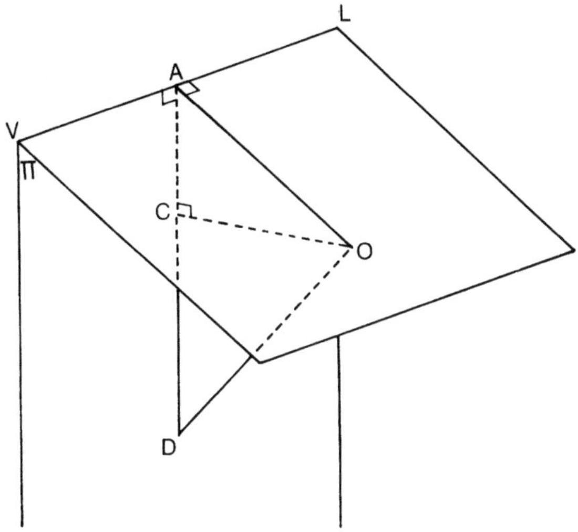

Figure 25. An illustration to Taylor's Problem XIV.

Taylor found a simple construction and gave a very concise proof of its validity. As earlier I shall supplement his arguments with some remarks. Thus let, in Figure 25, the line VL be the given vanishing line and C the center of the picture; since the *distance* is also given the position of the eye point, O, is known. Required is the vanishing point, D, of the normals to the plane OVL. By definition D is the point where the normal at O to plane OVL meets the picture plane, so let us analyze how that point can be constructed from the data.

Let OA be perpendicular to VL; then VL is orthogonal to OD as well as the AO and therefore a normal to plane AOD, in particular, it is perpendicular to AD. Moreover, the line AC is perpendicular to VL—a result Taylor stated in Theorem 1 of *New Principles* referring to Euclid's *Elements* XI,11. (It can also be established by noticing that VL is a normal to plane ACO.) As the lines AC and AD both lie in the picture plane and are normals to VL they coincide, hence D may be characterized as the point of intersection of AC and the normal at O to AO in the plane AOC. The point D will remain fixed if we rotate the triangle AOD around AC into the picture plane; this explains Taylor's construction when AB is the given vanishing line (Figure 26):

> Draw CA perpendicular to AB and CO parallel to it, and equal to the Distance of the Picture. Draw AO and OD perpendicular to it, cutting CA in D; which will be the Vanishing Point sought. [p. 198 = Taylor, 1719, p. 38]

In a note Taylor pointed out that the point D is determined on AC by the relation

$$AC : AO = AO : AD. \tag{6.1}$$

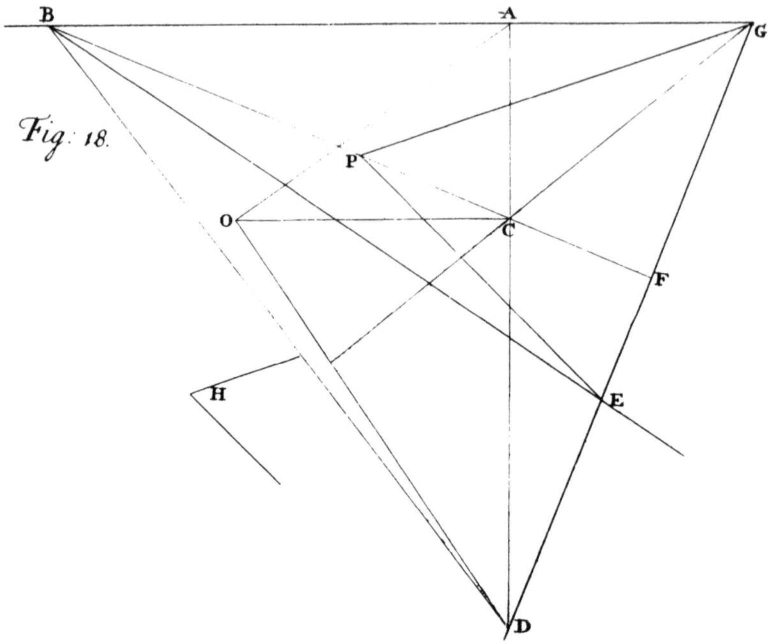

Figure 26. *C* is the center of this picture, and its *distance* is *CO*. The diagram shows—among other things—how to find the vanishing point *D* of normals to planes that have *AB* as their vanishing line. *New Principles*, Figure 18 (p. 237).

More remarkable, however, are some notes where he used the concept of points at infinity (p. 198). Thus he noticed that when the center of the picture *C* lies on the vanishing line (*VL* in Figure 25 and *AB* in Figure 26) the outcome of the construction is that the point *D* "will be infinitely distant." He interpreted this as meaning that the images of the normals to the plane *OVL* (Figure 25) will be the set of parallel lines perpendicular to *VL*. He found that this result is in accordance with the fact that when *C* lies on *VL* the plane *OVL* is perpendicular to the picture plane, and hence its normals are parallel to the picture plane and will be depicted in lines parallel to the originals. Similarly, he remarked that in the case where the original plane is parallel to the picture plane "the Distance *CA* will be infinite, and consequently *OA* will be parallel to *CA*, and *OD* will coincide with *OC*, making the point *D* to fall into the Center of the Picture *C*" (pp. 198–199).

Based on the construction of Problem XIV Taylor made several other observations of which we shall need the following in the next section (Figure 25). Given the vanishing line *VL* of a plane, and the vanishing point *D* of lines perpendicular to the plane, and let *DA* be the normal from the vanishing point to the vanishing line: then

(1°) the center of the picture *C* lies on *DA* (p. 201), and

(2°) the angle between the lines from the eye point to D and A is a right angle (*ibid*).

Furthermore, the considerations of Problem XIV lead straightforwardly to the solution of Taylor's Problem XV which is the perspective version of Problem 2 considered above (p. 34). Here we assume (Figure 25) that the vanishing point D, the center of the picture C, and the *distance* of the picture CO are given. We search for the vanishing line of the planes perpendicular to the lines whose vanishing point is D. The solution is to determine A on CD by the relation (6.1)—which can be done geometrically—the line in the picture plane that is normal to CD at A is then the required vanishing line.

Building upon Problems XIV and XV Taylor solved the perspective versions of Problems 3 and 4; I shall only deal with the latter which Taylor formulated in the following manner:

PROBLEM XVII

Having given the Center and the Distance of the Picture, and the Vanishing Point of the common Intersection of two Planes that are inclined to one another in a given Angle, and the Vanishing Line of one of them; to find the Vanishing Line of the other of them. [p. 201 = Taylor, 1719, p. 41]

Before presenting Taylor's solution I find it convenient to mention two useful results which Taylor derived as immediate consequences of the definitions of vanishing points and lines:

(i°) The vanishing line of a plane contains all the vanishing points of lines lying in the plane (cf. Theorem VII, p. 178). This means, in particular, that the vanishing line is determined by two vanishing points of lines lying in the plane.

(ii°) The vanishing point of the line of intersection of two planes is the point of intersection of the vanishing lines of the planes (cf. Corol. 2, p. 179).

Let then BG (Figure 26) be the given vanishing line in Problem XVII, B the given vanishing point, and H the given angle. It is required to determine the vanishing line of the set of parallel planes which make the angle H with the planes having the vanishing line BG and which intersect these in the set of parallel lines whose vanishing point is B. In Figure 27 I have attempted to illustrate the situation by representing the parallel planes and lines by the ones which pass through the eye point O. The line l has the vanishing point B. Using Problem XV Taylor determined the vanishing line DG of the planes perpendicular to l. The point G where DG meets the given vanishing line is according to (ii°) the vanishing point of the intersections of the planes perpendicular to l and the planes whose vanishing line is BG; in Figure 27 m is the line from this set which passes through O. Taylor proceeded by employing Problem XI (p. 27) to construct the vanishing point E on GD for the lines making the angle H with the lines whose vanishing point is G; in Figure 27 n makes the angle H with m. Finally Taylor claimed that BE is the required line.

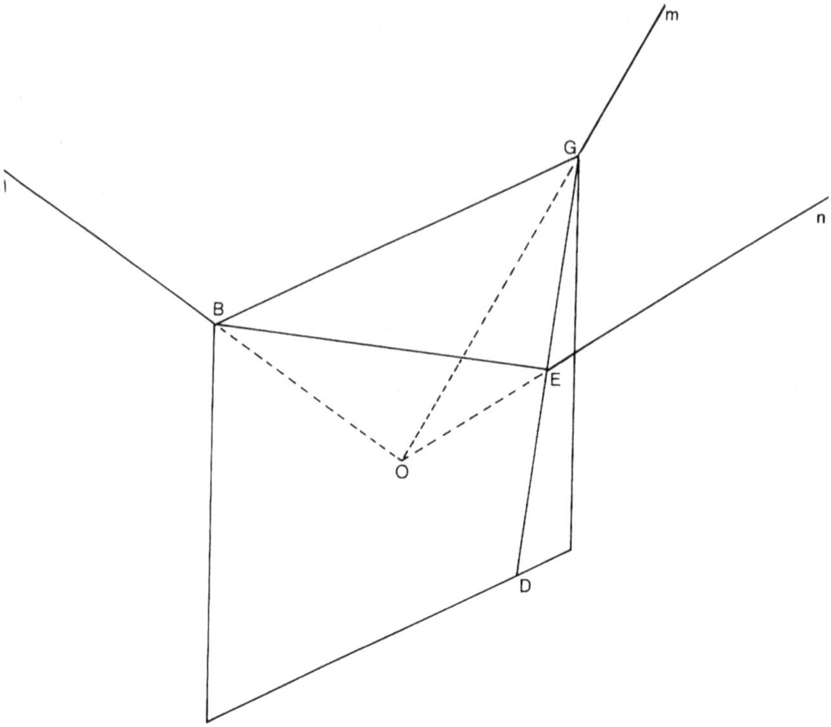

Figure 27

That this is the case may be realized in the following way (Figure 27). The plane (m, n) has the vanishing line GE (i°) which is the same as GD, hence the plane is normal to l—that is, to the intersection of the planes (l, m) and (l, n). This means that the angle between the latter planes is equal to the angle H between m and n. From this it follows that the required vanishing line is that of the plane (l, n) which is indeed BE (ii°).

To illustrate how Taylor intended his theory to be used for direct constructions I shall sketch how in *Linear Perspective* he threw a regular tetrahedron into perspective, or to be more precise how he solved the following problem (Figure 28) Given the image of the side HI of a regular tetrahedron, the vanishing line of the face HIK, the center and the *distance* of the picture; complete the image of $HIKQ$ (p. 100). First Taylor made the image of triangle HIK by drawing the lines HK and IK that perspectively form the angle 60° with HI. Then he used the fact that the angle between two faces of a regular tetrahedron is known and found the vanishing line of the face HIQ according to Problem XVII above (which occurs as Problem 7 in *Linear Perspective*). Thereafter he found the image of the triangle HIQ in the same way as the image of HIK was found in the first step, and this finished the perspective tetrahedron.

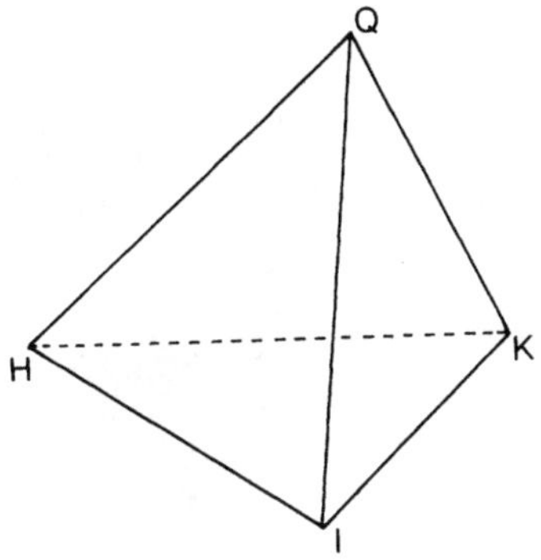

Figure 28

New Principles contains, as mentioned, a general description of the method just presented, but no examples of its application (p. 203). Instead, we find an example of another method which was not mentioned in *Linear Perspective* and thus probably was invented after 1715. This method is fascinating because it combines the traditional use of a plan and elevation with the approach of direct constructions. The idea of the method is to put a plane as well as an elevation of an object in perspective and to obtain the final image by a composition (p. 203). To illustrate how much theoretical insight the method requires I shall paraphrase Taylor's example of throwing a regular dodecahedron into perspective (Figure 29). He assumed that the center of the picture H, the distance of the picture OH, the perspective image AB of one of the sides of the dodecahedron, and the vanishing line FG (with center F) of the original of the face $ABCDE$ are given—he furthermore assumed that FG and AB are parallel.

Taylor's first step was to construct the perspective image of a plan of the dodecahedron and as plane of projection he chose one—which I shall denote α—perspectively parallel to $ABCDE$. An orthogonal projection of the dodecahedron into α in the picture plane becomes a central projection from the point, I, which is the vanishing point of the lines orthogonal to α (I is found on the normal to FG at F according to Problem XIV). Since AB is parallel to the vanishing line FG of α, the perspective image ab of the projection of the original of AB into α is also parallel to FG; ab can then be chosen as one of the line segments which have their end points on IA and IB and are parallel to AB. Having decided upon the position of ab Taylor completed the construction of the perspective plan of the dodecahedron—leaving the details to the readers.[13]

The next step was to construct the perspective elevation; as the plane of elevation Taylor chose one that has FO as its vanishing line. Let the plane of

[13] In *Linear Perspective* Taylor presented a construction of the plane and elevation of the dodecahedron (pp. 101–102, cf. Note 30, p. 138). The plan consists of two regular decagons having a known ratio between their sides. Hence its perspective image can be constructed from the side ab by employing Problem XIII (p. 33).

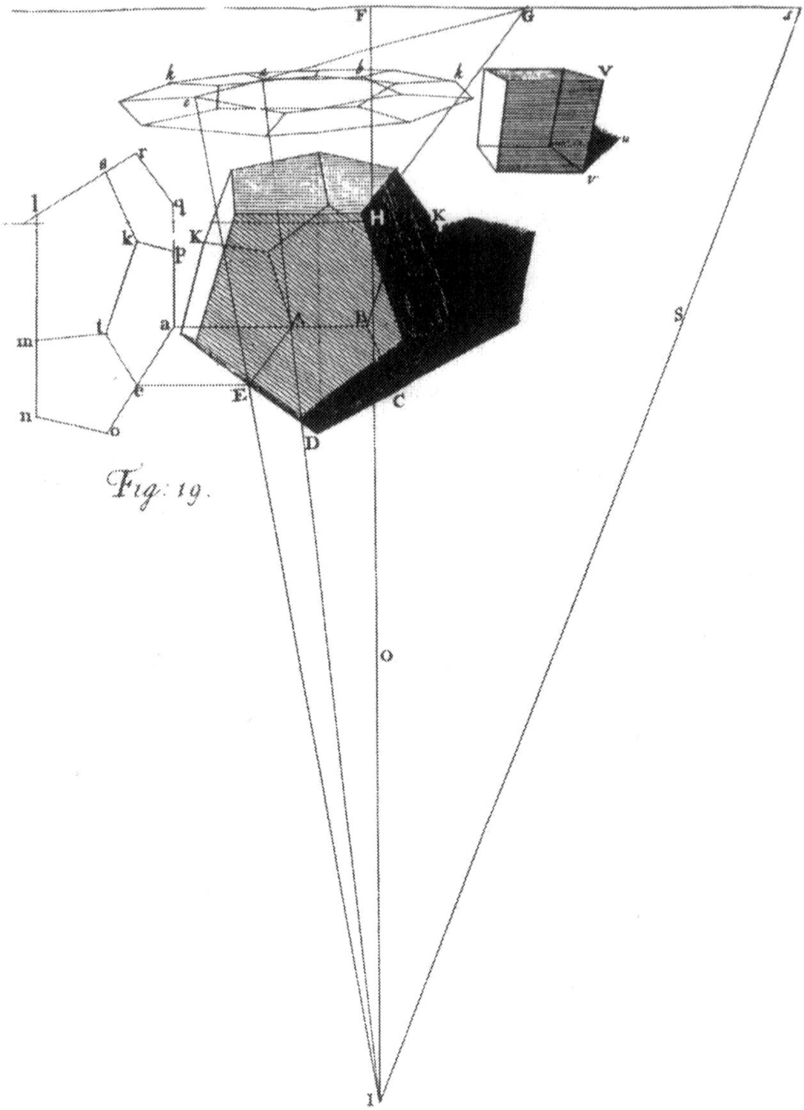

Figure 29. Taylor's construction of the perspective image of a dodecahedron lying on the face *ABCDE*. This face is perspectively parallel to the top face and has the vanishing line *FG* which is parallel to the side *AB*. The plan of the dodecahedron is constructed on a plane perspectively parallel to *ABCDE*, and its elevation on a plane perspectively parallel to the plane which is perspectively perpendicular to *ABCDE* and contains *D* and the perspective midpoint of *AB*. *New Principles*, Figure 19 (p. 238). A white H has been added to make the point *H* visible.

elevation be denoted β. Since the center H of the picture lies on FO, lines perpendicular to β are depicted in the perspective plane as lines perpendicular to FO (a result Taylor established—as mentioned on page 36—in a note to Problem XIV). From this it follows that the orthogonal projection of the dodecahedron upon β in the perspective plane becomes a parallel projection with direction FG. To perform the construction of the perspective elevation Taylor needed one of its sides; he could not obtain this from AB since its projection is a point. He therefore made a direct construction of the point E of the perspective dodecahedron using the point e of the perspective plan and the following considerations. Since the originals of the lines AE and ae are parallel, the vanishing point, G, of ae is also the vanishing point of AE, hence the point E is obtained as the point of intersection of AG and Ie.

Taylor then proceeded by projecting AE into **ae** by using the fact that A**a** and E**e** are parallel to FG and that **ae** has the vanishing point F (because **ae** lies in the plane $ABCDE$, as well as in a plane with vanishing line FO, cf. (ii°), p. 37). Upon **ae** he completed the perspective elevation of the dodecahedron (cf. p. 206; in Figure 1, p. 246, the nonperspective elevation is illustrated). Finally, he composed the perspective plan and elevation to one image by determining the points where the lines from I to the vertices of the plan meet the lines parallel to FG through the vertices of the elevation.

Theoretically, the method just described is both interesting and well founded; its practical value, however, is to be doubted because the construction is so complicated; what makes it particularly complicated is that Taylor —unlike other authors—allows an angle different from 90° between the picture plane and the bottom face of the dodecahedron. The example indicates that Taylor was less interested in presenting the various steps in a construction than in pursuing their theoretical background. This impression is confirmed by the last examples of *New Principles* in which he only discussed which part of his theory he had applied. The attitude of expecting the readers to realize what precisely the problems were and to be able to carry out the constructions themselves is also reflected in Taylor's treatment of shadows.

Some of Taylor's predecessors had discussed—but in general only briefly —the problem of constructing shadows in a perspective picture. Taylor kept to this tradition of only spending a few words on shadows, thereby missing an opportunity to stress how powerful his method of direct constructions was. That it works elegantly for shadows I shall illustrate by describing Taylor's construction of the shadow cast by the cube in Figure 29. The source of light is supposed to be the sun and the light rays are considered to be parallel lines having S as vanishing point. The shadow is assumed to be cast on the plane that contains the bottom face of the cube which is parallel to the face $ABCDE$ of the dodecahedron, and hence has the vanishing line FG. The point I was earlier determined as the vanishing point of lines perpendicular to $ABCDE$, I therefore is also the vanishing point of the side Vv of the cube.

Constructing a shadow is a question of determining projections in the perspective plane; this can be done by finding those lines and points that are

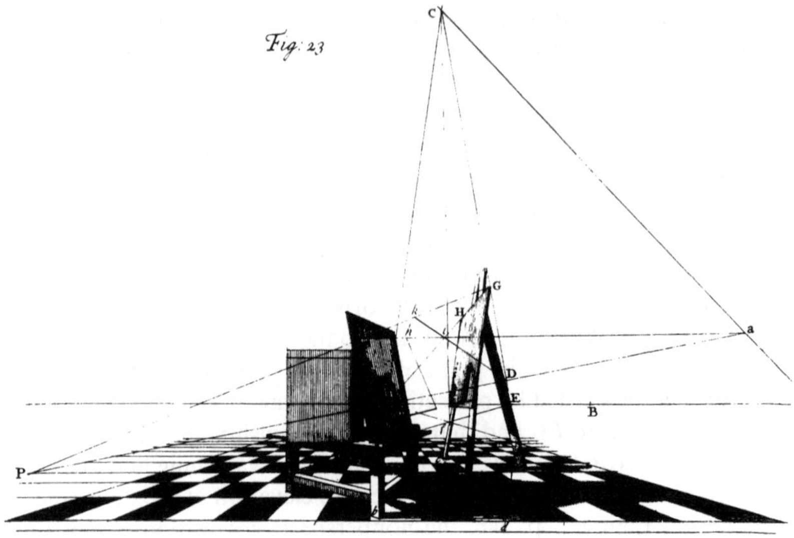

Figure 30. A copy of Figure 23 from *New Principles* (p. 242). As it is very difficult to distinguish the various letters the main lines of the diagram are reproduced below.

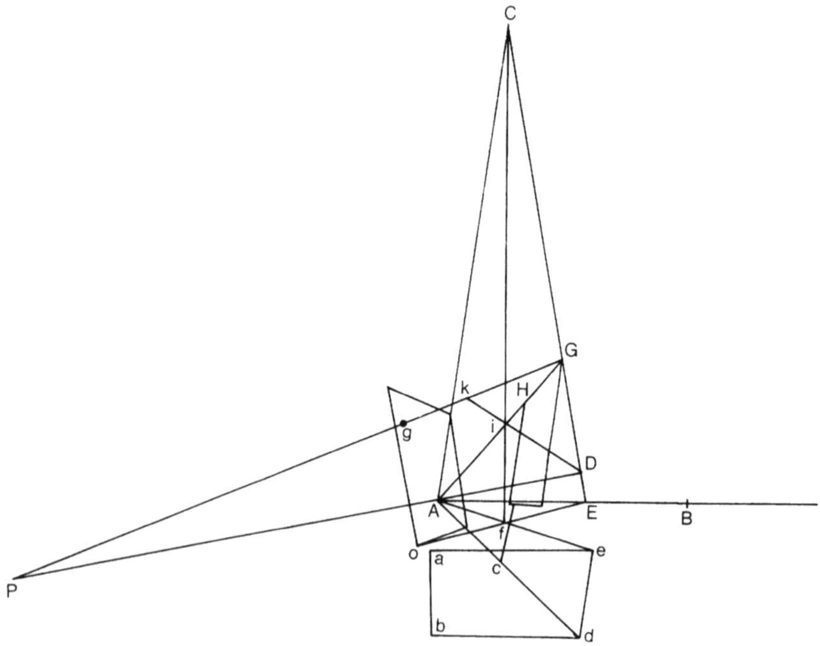

intersections of the relevant planes and lines. For example, to find the shadow of the line *Vv* we have to find the image, *vs*, of the line in which the plane (α) upon which the shadow is cast meets the plane (μ) containing all the rays of light passing through the points on *Vv*. We know that *v* lies on this line, hence it is determined if we can find its vanishing point. The latter is the point of intersection of the vanishing lines of α and μ ((ii°), p. 37). The first is given as *FG*, and since *I* and *S* are vanishing points for two directions of lines in μ, *IS* is the vanishing line of μ ((i°), p. 37); thus the point of intersection *s* of *FG* and *IS* is the vanishing point of the trace between α and μ, and *vs* is the shadow of *Vv*. To find out where the shadow of the point *V* falls Taylor constructed the point of intersection, *u*, of *vs* and *VS*.

As the last example of Taylor's direct construction I shall mention what was also his last example in *New Principles*. In the illustration belonging to it (Figure 30), we see among other things the reflection of an oblique painting in an oblique mirror. This is obviously complicated, so complicated that although Taylor kept to his usual concise style he needed two pages to describe how the reflection of the vertex *G* of the painting is determined (pp. 212–213). It is given that *A* is the center of the picture, *AB* the *distance* of the picture, *AC* the vanishing line of the painting, and *CD* that of the mirror. Using Problem XIV (p. 34) Taylor found the vanishing point, *P*, of the normals to the mirror; furthermore, juggling with vanishing points and lines he found the perspectively orthogonal projection, *k*,[14] of the point *G* on the plane of the mirror.

---

[14] It is a long process to construct the point *k*. To make Taylor's arguments more comprehensible I shall specify which planes and which traces between planes he considered; I denote planes by Greek letters and use the symbol $\cap$ for traces. Determinations of traces are an especially important part of Taylor's construction. One should therefore keep the result (ii°) on page 37 in mind. The first two planes to be considered are (Figure 31):

μ—the plane of the mirror (with vanishing line *CD*),
π—the plane of the painting (with vanishing line *AC*).

For his construction Taylor needed the trace $\mu \cap \pi$ which is rather complicated to find because the data do not immediately lead to this line. It has vanishing point *C* (ii°), but to find another point on the line Taylor had to take several steps. First he used:

τ—the plane of the table (with vanishing line *AB*),

and wanted to get a second point on $\mu \cap \pi$ as the point of intersection of $\mu \cap \tau$ and $\pi \cap \tau$. The determination of the latter is, however, not obvious either; the vanishing point of $\pi \cap \tau$ is *A* (ii°), but to find a second point on $\pi \cap \tau$ requires some further search. Thus Taylor introduced

γ—the ground plane or floor (with vanishing line *AB*),

and

η—the plane normal to γ which passes through the edge *ba* of one of the legs of the table and is parallel to the perspective plane.

Taylor's idea was now to find a second point on $\pi \cap \tau$ as the point of intersection of $\pi \cap \eta$ and $\tau \cap \eta$. To find the first line he had to make yet another detour, namely, via $\pi \cap \gamma$ and $\eta \cap \gamma$. The trace $\pi \cap \gamma$ has vanishing point *A* and contains the point *c* which is given as lying in π as well as in γ, hence $\pi \cap \gamma$ is the line *Ac*. The trace $\eta \cap \gamma$ contains the point *b*, therefore—since η is

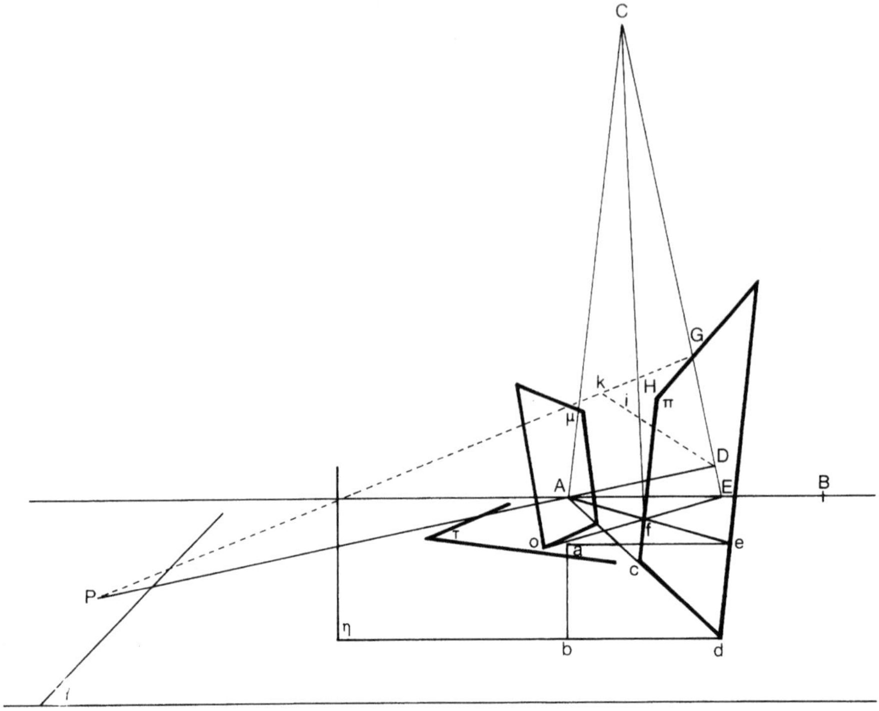

Figure 31

parallel to the perspective plane and γ has vanishing line AB—$\eta \cap \gamma$ is the line, bd, through b parallel to AB. By finding the point of intersection, d, of Ac ($\pi \cap \gamma$) and bd ($\eta \cap \gamma$) we have obtained a point on $\pi \cap \eta$; using the fact that η is parallel to the perspective plane and π has vanishing line AC we conclude that $\pi \cap \eta$ is the line, de, through d parallel to AC. The other required trace $\tau \cap \eta$ is similarly found as the line, ae, through a parallel to AB.

We now have a second point on $\pi \cap \tau$, namely, the point of intersection, e, of de ($\pi \cap \eta$) and ae ($\tau \cap \eta$), thus $\pi \cap \tau$ is Ae. The other trace $\mu \cap \tau$ is easier to find, because it is given that the point o lies in μ as well as in τ, thus $\mu \cap \tau$ is the line Eo (E is the point of intersection of CD and AB). The determination of a second point on $\mu \cap \pi$ is then eventually finished: it is the point of intersection, f, of Eo ($\mu \cap \tau$) and Ae ($\pi \cap \tau$), and $\mu \cap \pi$ is therefore the line Cf.

Taylor proceeded by determining the orthogonal projection of the line GH (containing a side of the painting) upon μ, that is, the trace between μ and

α—the normal plane to μ containing GH.

Let us first determine α's vanishing line. It contains the vanishing point, P, of normals to μ which is constructed according to Problem XIV, moreover, it contains the vanishing point of GH which is given as A, hence the vanishing line of α is AP. Thus the trace $\alpha \cap \mu$ has the vanishing point D (the point of intersection of AP and CD). As a second point on $\alpha \cap \mu$ Taylor chose the point of intersection of GH and μ; precisely the determination of this point made it relevant that he knew $\mu \cap \pi$, because the required point is the point of intersection, i, of GH and $\mu \cap \pi$ (Cf), thus the line GH is projected onto the line Di. Finally, the point k, the projection of G onto μ, can be determined as the point of intersection of Di and PG (the normal to μ through G).

Transforming the reflection situation from Euclidean space to the perspective plane (Figure 30), Taylor then determined the reflection of G, g, by the requirement that Gk and kg shall be perspectively equal. In *Linear Perspective*, Taylor had a small section called "Of finding the Representations of the Reflections of Figures on polish'd Planes" (p. 109), where he used perspective equality to find the reflections of points and lines and to find the vanishing line of the reflection of a plane when various things are given; he did not include these problems in *New Principles*.

This last example clearly illustrates how elegantly Taylor worked directly in the perspective plane and that his theory contains all the insights necessary for a perspective geometry. However, the final decisive step of providing the picture plane with its own geometry was—as mentioned earlier—left to Lambert.

7. Inverse Problems of Perspective

Standing in front of a perspective picture it is natural to ask, "Where shall I put my eye to perceive what the artist had in mind?". This is an inverse problem of perspective in which knowledge of the original configuration is sought from a perspective picture. Formulated generally, an inverse problem of perspective is obviously indeterminate, but it becomes soluble if some assumptions are made. Let us, for instance, suppose that a perspective image in a vertical picture contains a chequered floor where one set of lines converges towards one point and the other set of lines is horizontal; then the assumption that the tiles of the floor are images of squares enables us to conclude that the point of convergence is the principal vanishing point of the picture and to determine the *distance* between the picture plane and the eye point, namely, as the distance between the principal vanishing point and the vanishing point of the diagonals of the squares.

Guidobaldo mentioned a few results concerning inverse problems of perspective but only as lemmas to usual perspective problems (Guidobaldo, 1600, pp. 110–112). Stevin made it a real topic of investigation and he seems to have been challenged by the idea of making as few assumptions as possible. Some of the seventeenth-century writers who succeeded Guidobaldo and Stevin also touched upon inverse problems of perspective, but they did not attempt to continue Stevin's more systematic approach. As far as I am aware Taylor was the first to take that up again; for instance, the two Dutch mathematicians Frans van Schooten and 's Gravesande, who in other respects followed Stevin's line, did not engage themselves in inverse problems of perspective. Whether Taylor had a source of inspiration for his study of inverse problems is unclear, but presumably he had come across some problems and from them he worked out a more coherent theory.

Taylor distinguished between two types of inverse problems and described them in the following manner:

> ... to find out what Point the Picture is to be seen from, or having that given to find what the Figures are which are described on the Picture. [Taylor, 1715_2, p. 303]

In *Linear Perspective* (pp. 112–116) Taylor treated four problems, and he repeated these and added another three in *New Principles.* In determining the eye point he started with a problem where the following is given: a perspective triangle, the vanishing line, and the angles of the original triangle (Problem XXI, p. 216). He then proceeded to a perspective quadrilateral (Problem XXII, p. 217), but in this case he assumed that only the "Species"—i.e. the angles and the ratios between the sides of the original quadrilateral—are given. From these data he determined the vanishing line of the original quadrilateral. His tool for this was the result that when the images *A*, *B*, and *C* of three points on a line and the ratio $A_0B_0 : B_0C_0$ between the original line segments are given, then the vanishing point of the original line can be determined (Problem XIX, p. 215, which is the inverse of Problem III discussed on p. 27). Applying this result to the two diagonals of the quadrilateral Taylor found its vanishing line.

In the case where the eye point is given Taylor treated the general problem of determining the original of a perspective polygon, when the intersection and the vanishing line of the plane of the original polygon are given (Problem IX, p. 190). He also dealt with the situation where only the original length of a perspective line segment is required—the intersection and the vanishing line of a plane containing the original line segment being given (Problem X, p. 191). Furthermore, he showed—using Problem XI (p. 30)—how the "Species" of a triangle can be found when its vanishing line is given (Problem XX, p. 215).

Taylor closed his treatment of inverse problems with an example concerning a three-dimensional figure where the eye point as well as the "Species" have to be found:

> PROBLEM XXIII
>
> Having given the Projection of a Right-angled Parallelopiped; to find the Center and the Distance of the Picture, and the Species of the Original Figure. [p. 217 = Taylor, 1719, p. 57]

Let *ABCDEFG* (Figure 32) be the given perspective image of a right-angled parallelepiped. To find the eye point—or, equivalently, the center and the *distance* of the picture—Taylor built upon results obtained in connection with Problem XIV. First he found the vanishing points *H*, *I*, and *K* of the three sets of parallel sides of the parallelepiped. (For instance, *H* is the point of intersection of *AB* and *DC*.) The lines *HI*, *HK* and *KI* are then the vanishing lines of the three sets of parallel faces ((i°), p. 37). Since *K* is the vanishing point of lines perpendicular to planes which have *HI* as their vanishing line, it follows from (1°), page 36, that the center of the picture, *S*, lies on the normal *KL* to *HI*. Similarly, *S* lies on the normal *HM* to *KI*, and hence is the point of intersection of *KL* and *HM*. In connection with Problem XIV it was also

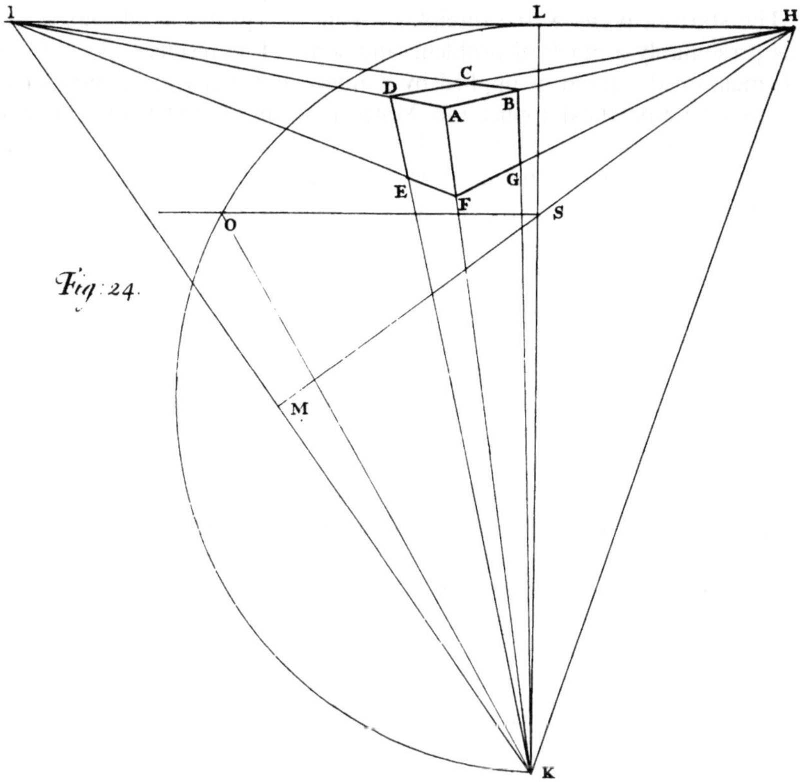

Figure 32. An inverse problem of perspective. *New Principles*, Figure 24 (p. 243).

observed that the angle between the lines from the eye point to K and L is a right angle ((2°), p. 37); this implies that the distance of the picture is equal to OS when O is the point where the circle with diameter KL meets the normal to KL at S. Taylor now knew the center and the *distance* of the picture and the vanishing lines of the faces of the parallelepiped; he then proceeded by using the just-mentioned result concerning a triangle whose vanishing line is given to find the ratios between the sides of the parallelepiped.

In a corollary Taylor looked at the situation where K is a point at infinity. He pointed out that S will then be a point on HI whose position cannot be determined unless the ratio between the originals of AB and BC are known. Inversely, if it is only required that the perspective image shall represent a right-angled parallelepiped the eye point may be any point on the perimeter of a circle that is perpendicular to the picture plane and has HI as diameter. Taylor found this a useful result stating:

> This I leave as a hint that may be useful to the Painters of Scenes in Theatres. [p. 218 = Taylor, 1719, p. 58]

The starting point of the discipline of inverse problems of perspective was presumably a practical problem, and some of its results—like the one just mentioned—could be applied by painters, but many of its results had no immediate practical application. Mathematicians like Stevin and Taylor were not primarily looking for such, but wanted to gain more theoretical insight. Later, when photogrammetry had been invented, Stevin's, Taylor's, and others' investigations turned out to be very useful.

8. The Appendices of *New Principles*

New Principles ends with two appendices, one dealing with a practical method of making perspective drawings on a curved surface and one treating the subject of mixing colors. None of these themes are treated in *Linear Perspective*, but in his account of this book in the *Philosophical Transactions*, Taylor presented the basic idea of the first appendix (Taylor, 1715_2, pp. 303–304). This was to find projections of lines as shadows of threads. More specifically, he suggested that the problem of finding the perspective image of a line on a curved surface—i.e. finding the intersection of the surface and the plane determined by the line and the eye point—should be solved in the following way: A thread is stretched out through the eye point parallel to the given line and a light is placed so that the shadow of the thread coincides with the line. This means that the light rays between the thread and the given line compose the plane whose intersection with the curved surface is required. Hence this intersection can be determined as the shadow of the thread on the curved surface.

Taylor repeated this idea in *New Principles* and adjusted the method to the fact that in most situations the curved surface will—seen from the eye point—cover the original line. He thus suggested that the projections of two points of the original line on the curved surface should be found by "some proper Method" (p. 220); the image of the entire line can then be found by placing the light so that the shadow of the thread passes through these two points. Otherwise, Taylor did not elaborate much on the method, stating that:

> I shall not enlarge upon this Method, not having had an opportunity of putting it in practice; for which reason I only propose it as a Hint, which I leave to be further consider'd of by the Curious. [p. 221 = Taylor, 1719, p. 61]

The second appendix is called "A New Theory for mixing of Colours, taken from Sir Isaac Newton's *Opticks*," and in it Taylor presented what later was called Newton's color circle. This served as a model for explaining a mixing of prismatic colors (Figure 33) according to which the colors to be mixed are represented by weights that are placed along the arcs of the circle and are proportional to the numbers of rays of the particular colors: the color resulting from their mixture will then be represented by the center of gravity of the weights.

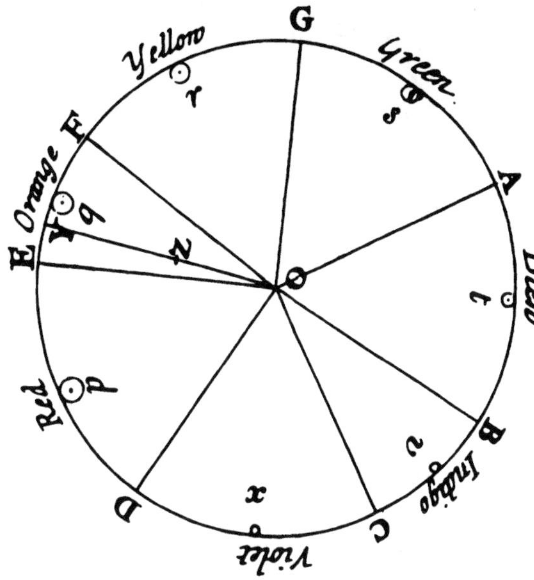

Figure 33. Newton's color circle; the circumference is divided "into seven parts *DE*, *EF*, *FG*, *GA*, *AB*, *BC*, *CD* proportional to the seven musical Tones or Interval of the eight Sounds Sol, la, fa, sol, la, mi, fa, sol, contained in an Eight, that is, proportional to the numbers 1/9, 1/16, 1/10, 1/9, 1/10, 1/16, 1/9." (Newton, 1704, p. 114 and Figure 11.) The seven parts represent the principal spectral colors.

In describing Newton's circle Taylor made some minor technical changes and the following result emerged (Figure 34): The circumference of a circle with center O is divided according to Newton's description (Figure 33). All points on a radius like OS represent the same color, while a point on the circumference represents what Newton called the most intense—and Taylor the simplest or cleanest—form of the color; the nearer a point on OS is to O the more compound or broken is the color it represents; the point O itself represents white.

To illustrate how the circle can be used for finding the result of mixing colors, Taylor gave the following example: Suppose that the colors represented by P and Q, respectively, are mixed in the ratio 2 : 3, Taylor then found "the Center of Gravity **3** of the Points P and Q" (p. 225) (that is, he determined the point **3** on PQ so that $P\mathbf{3} : \mathbf{3}Q = 3:2$) and claimed that the point **3** represents the mixed color. The endpoint **1** of the radius through **3** told Taylor which color he had obtained, and the position of the point **3** on $O\mathbf{1}$ how broken the color was. The procedure can be continued, thus if one adds to the mixture just found five parts of the color represented by R the midpoint, r, of $\mathbf{3}R$ represents the new mixture.

With the use of vector notation Taylor's application of Newton's theory of color mixing can be described as follows (Figure 35): Let A and B represent two colors inside the color circle. If these colors are mixed in the ratio $m:n$

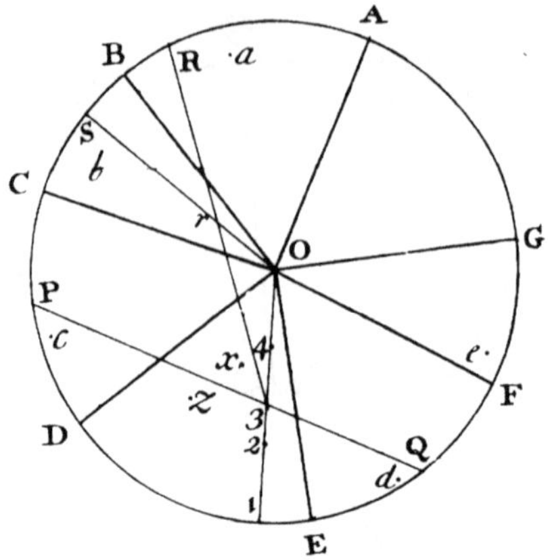

Figure 34. Taylor's version of Newton's color circle; the arc AB contains the various reds, BC the oranges, etc. (cf. Figure 33). *New Principles*, Figure 25 (p. 243).

the resulting color is represented by a point M determined by

$$\overrightarrow{OM} = \overrightarrow{OA} + \overrightarrow{AM} = \overrightarrow{OA} + (n/(m+n))\overrightarrow{AB}$$
$$= (1/(m+n))(m\overrightarrow{OA} + n\overrightarrow{OB}). \quad (8.1)$$

This formula shows that the theory is consistent in the sense that it does not matter in which order several colors are mixed. Newton—and Taylor with him—took this for granted, or rather they implicitly based this result on their experiences in working with centers of gravity. Taylor also employed the color circle inversely and showed how one can obtain a given color represented by a point in the color circle by decomposing it into other colors according to a procedure that corresponds to formula (8.1). Taylor, furthermore, used the color circle to argue that for mixing colors it is best to have a sample of simple colors, for instance, if one only possesses the colors represented by a, b, c, d, and e (Figure 34), it is not possible to obtain a color represented by a point outside the polygon $abcde$.

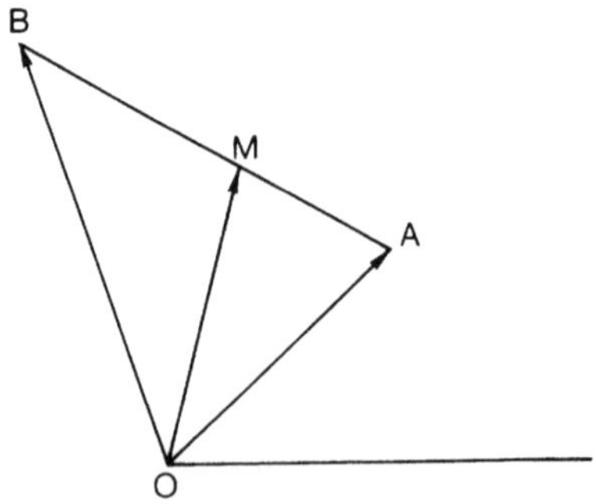

Figure 35

In principle, Newton's rules apply only for mixing light rays; Newton did not say much about mixing pigments in Proposition VI of *Opticks* where he explained his color circle, but at one point dealing with the ratio—he said proportion—in which an orange and a white should be mixed, he added:

> this proportion being not of the quantities of mixed orange and white powders, but of quantities of the lights reflected from them. [Newton, 1704, p. 117]

In general, Newton thought that the mixing of pigments by and large follows the same rules as the mixing of light rays, and therefore supposed—as pointed out by Alan Shapiro—that his color circle was valid for pigment mixing.[15]

Taylor was well aware of the fact that Newton's rules had been developed for light rays, but in accordance with the above-quoted sentence from Newton he applied them also for pigments with the following modification:

> if several artificial Colours were to be mix'd according to these Rules, and some of them are darker than others, there must be a greater Proportion used of the darker Materials, to produce the Hue [color] proposed, because they reflect fewer Rays of Light in proportion to their Quantities; and a lesser Proportion must be used of the lighter Materials, because they reflect a greater Quantity of Light. [p. 227 = Taylor, 1717, p. 67]

Unfortunately, I have not been able to find out to what extent later practitioners, building upon Taylor's theory of linear perspective, also utilized the theory of mixing colors. But the fact that Taylor chose to present this theory shows how concerned he was about having the art of painting based on scientific methods. For linear perspective he applied geometry in a more general form than was usual in textbooks and for mixing colors he applied the newest theory of light and color.

9. Taylor and the History of Linear Perspective in England before 1800

Unlike the Continent, England had no tradition of literature on perspective when Taylor engaged himself in the subject. If we compute the number of Italians, Frenchmen, Germans, and Dutchmen who wrote on perspective in some detail before 1700 the result will be at least sixty.[16] A similar count in England gives the number one. The person in question is the hydrographer

[15] In wondering about the application of Newton's color circle I have found much inspiration in a paper by Alan Shapiro, where he shows that although Newton in other connections demonstrated that he understood subtractive mixing, he had not realized that pigments mix subtractively, but thought that his rules for additive mixing of light rays applied to pigments as well (Shapiro, forthcoming).

[16] A bibliography of books and tracts on perspective up to 1825 is included in Jones, 1947, one up to 1600 in Schüling, 1973, and one up to Monge in Vagnetti, 1979.

Joseph Moxon who in 1670 published *Practical Perspective; or Perspective Made Easie*. His choice of title seems to have been rather influential, because the eighteenth century witnessed the publication of several books whose titles claimed that they had made perspective easy.[17] The content of Moxon's book, however, did not leave any noticeable traces.

Before 1710 there were a few more possibilities of reading about perspective in English—namely in translations of some of the famous continental works, such as Sebastian Serlio's work on architecture (translated 1611), Jean Dubreuil's *La perspective practique*, better known as "The Jesuit's perspective" (translated 1672), and Andrea Pozzo's *Perspectiva pictorum et architectorum* (translated 1707). From these books the readers could learn the rules of perspective, but they would find no scientific explanation of these rules.

At the beginning of the second decade of the eighteenth century two more books on perspective were published in English; they both treated the practice as well as the theory of perspective. The first was a translation, appearing in 1710,[18] of Bernard Lamy's *Traité de perspective* from 1701. The second was composed in English by Humphry Ditton; he had earlier published an exposition of Isaac Newton's *Principia*, and presumably through Newton's influence he got a position as master at the New Mathematical School at Christ's Hospital (which position only existed until Ditton's death in 1715). At the school, Ditton taught perspective and in 1712 he published the textbook *A Treatise on Perspective Demonstrative and Practical*. The quality of the book is above average, nevertheless, Ditton had apparently very few readers; I suspect—but cannot prove—that Taylor was among them.

In total there existed less than ten English treatises on perspective when Taylor published his *Linear Perspective* in 1715. His contribution was a tremendous improvement on the theory of perspective, but the book was mathematically too advanced for the general public and does not seem to have received much attention in England.

On the continent *Linear Perspective* was noticed, but the continental mathematicians were much more concerned about Taylor's other publication from 1715, *Methodus incrementorum*, to which they gave very unfavorable reviews in *Acta Eruditorum* as well as in private correspondence; in the latter sometimes with some ambiguity so that it is unclear which of Taylor's two books are meant. Johann Bernoulli was one of Taylor's continental colleagues who was interested in perspective. He had praised 's Gravesande's *Essai de perspective* highly,[19] whereas he is reported to have been far from enthusiastic

[17] For instance, Bardwell, 1756, Ferguson, 1775, Halfpenny, 1731, Kirby, 1754, and Lamy, 1710.

[18] Some authors mention an English edition of 1702 but I have not been able to verify this.

[19] The 20 March 1714 Johann Bernoulli wrote to 's Gravesande:

> "Je vous supplie de l'accepter comme venant d'une Personne qui a beaucoup d'égard & de considération pour votre mérite et savoir dans les Mathématiques, dont j'ai vu une preuve suffisante par l'excellent Traité sur la Perspective que vous avez publié ....

about *Linear Perspective.* According to Taylor's grandson, Young, Bernoulli wrote that it was "abstruse to all and unintelligible to artists for whom it was more especially written" (Taylor, 1793, p. 29). This sentence is often quoted, its authenticity is, however, doubtful.[20] But anyway, considering that Bernoulli and Taylor were some of the leading figures in the bitter cross-channel fight about the priority of the invention and various applications of the calculus, and considering that Bernoulli had a mordant style, it is not surprising if Bernoulli had expressed himself negatively about *Linear Perspective.*

Moreover, the point that Taylor's work was recondite was right. It seems, in general, to have been a problem for Taylor to express his ideas clearly: his main work *Methodus incrementorum* was characterized as obscure, not only by his opponents, but also by his friends (Feigenbaum, 1986). Presumably, Taylor himself realized that his ideas on perspective could be presented better than he had done in *Linear Perspective*, and therefore decided to write *New Principles*; at least he admitted in the preface of the latter that he was aware of some criticism:

> I find that many People object to the first Edition that I gave of these Principles in the little Book entitled, *Linear Perspective*, & c. because they see no Examples in it, no curious Descriptions of Figures, which other Books of Perspective are commonly so full of, and seeing nothing in it but simple Geometrical Schemes, they apprehend it to be dry and unentertaining, and so are loth to give themselves the trouble to read it. To satisfy these nice Persons in some measure, I have made the Schemes in this Book something more ornamental .... [pp. 152–153 = Taylor, 1719, pp. vi–vii]

The essential point of criticism—that Linear Perspective was too concise—he did, however, not take:

> It would have been easy to have multiplied Examples, and to have enlarged upon several things that I have only given Hints of, which may easily be pursued by those who have made themselves Masters of these Principles. Perhaps some People would have been better pleased with my Book, if I had done this: but I must take the freedom to tell them, that tho' it might have amused their Fancy something more by this means, it would not have been more instructive to them. [p. 153 = Taylor, 1719, p. vii]

The result of his efforts of once more trying to explain his ideas briefly may have disappointed him, for there is no evidence that *New Principles* at first got more attention than *Linear Perspective.* This is understandable, because

---

J'y ai trouvé plusieurs règles fort ingénieuses & très commodes pour la pratique que l'on ne trouve pas par tout ailleurs ...." ['s Gravesande, 1774, p. XI]

[20] Young claimed to be citing *Acta Eruditorum*; but in this journal Bernoulli did not express his views on *Linear Perspective.* Jones has suggested that the source for Young's quotation is a letter from Taylor to Monmort, where, however, Taylor is most likely referring to Bernoulli's opinions on *Methodus incrementorum* (Jones, 1951, Note 4).

New Principles is not essentially easier to read than *Linear Perspective*; in fact, the two books together constitute the mathematically most difficult pre-nineteenth-century literature on perspective.

At the time when *New Principles* was published the English interest in perspective was steadily growing and resulted in further translations of some of the important continental writers on perspective. In 1721 an English edition of some of Leonardo da Vinci's writings on perspective appeared, and in 1724 one of 's Gravesande's *Essai de perspective*, moreover, another English translation of Dubreuil's popular perspective came out in 1726 and went through several editions.

According to John Lodge Cowley, Taylor made a third attempt to make his ideas understood, but he died before he completed the last version of his work on perspective (Cowley, 1766, p. vii). Sadly enough it was only after his death in 1731 that Taylor's achievement made any mark, starting with John Hamilton's *Stereography or a Compleat Body of Perspective* from 1738. All through this book there are evidences of Taylor's influence which Hamilton fully acknowledged:

> ... in particular [I] have taken all such Assistance and Hints as were furnished me by Dr. Brook Taylor's two small Treatises on this Subject published some Years since, in which that learned Gentleman has, in a few Pages, made more Advances towards perfecting the Science, than all the Writers who went before him. [Hamilton, 1738, p. a.4]

Hamilton aimed at much more than Taylor, namely at a complete study of projections following—as he himself claimed—the line which Philippe de la Hire laid down in his work on conic sections. Hamilton spent 400 pages on his project and took up many interesting themes as projections of conic sections and of sets of harmonic points. He also investigated what could be called the curve of foreshortening: Let the distance between a point in the ground plane and the ground line be x and the distance between the perspective image of the point and the ground line be y; Hamilton then showed that the points (x, y) lie on a hyperbola—a result which among others Lambert also later noticed.

Hamilton addressed his book to readers with some mathematical knowledge, and for them he wrote a book which is remarkable because it unifies some of the continental ideas concerning synthetic geometry with Taylor' s approach to perspective. Hamilton maintained that:

> ... of late Years, no general Courses of Mathematicks have been esteemed compleat, without a particular Treatise on that Subject [Perspective]. [Hamilton, 1738, p. a.2]

Nevertheless, Hamilton's approach remained rather unusual. It was not perspective as a part of more advanced courses in mathematics, that characterized the eighteenth-century literature on the subject. Most of the English treatises on perspective published after 1740 are written for practitioners, irrespective of whether their authors were scientists or practitioners.

The practitioners' books are particularly interesting because they have a new style. Traditionally, practitioners had mainly composed manuals on perspective, that is, expositions explaining *how* to make perspective drawings; the readers who also wanted to know the *why* had to turn to books written by mathematicians. This practice was changed by the two rather well-known English painters, John Joshua Kirby and Joseph Highmore. They shared a vivid interest in perspective and a wish to bridge the gap between the practice and the theory of perspective; moreover, both of them found their inspiration for theoretical work on perspective in Taylor's *New Principles.*

Apparently Kirby and Highmore planned about the same time, and independently, to write books on perspective based on Taylor's theory. In the pamphlet *A Critical Examination* ... (1754) Highmore announced that he had been working on his book for some time. When in 1763 he eventually published it with the title *The Practice of Perspective on the Principles of Dr. Brook Taylor, written many years since* ... he had been preceded by Kirby at least twice and most likely three times. In 1754 Kirby's *Dr. Brook Taylor's Method of Perspective Made Easy, both in Theory and Practice* appeared, and seven years later followed *The Perspective of Architecture Deduced from the Principles of Dr. Brook Taylor.* In between, he saw Isaac Ware's translation of Lorenzo Sirigatti's *Practice of Perspective* described in the following way:

> The *best* Author that *ever* treated on Perspective is now translated from the Italian language into English. [Kirby, s.a.; Introduction]

This notice made him so upset that he wrote *Dr. Brook Taylor's Method of Perspective, compared with Examples lately published on this Subject, as Sirigatti's by Isaac Ware* (date of publication unknown, but according to De Morgan it was 1757). Kirby's first book seems to have been rather popular as it was reissued at least three times. The first edition was dedicated to William Hogarth, who also contributed to the book by making a festive frontispiece showing what can happen when artists do not know perspective (Figure 36).

Kirby's acknowledgment of Taylor's influence is similar to Hamilton's:

> I have entitled this Treatise DR. BROOK TAYLOR'S PERSPECTIVE, &c. out of Gratitude to that ingenious Author, for furnishing me with the Principles to build upon; and because his, though a very small Pamphlet is thought the most correct, concise and comprehensive Book upon the subject .... [Kirby, 1754, pp. i–ii]

He also had some reservations:

> But, not withstanding both these Treatises [*Linear Perspective* and *New Principles*] are so curious and useful, few have been able to understand his Schemes; and when they have understood them, have been as much puzzled in applying them to Practice. [Kirby, 1754, p. ii]

Highmore was of the same opinion:

> He [Taylor] has invented, and, in a very short compass, exhibited an universal theory; the truth, and excellence of which is acknowledged by all who have read,

Figure 36. William Hogarth's frontispiece to Kirby's *Dr. Brook Taylor's method of perspective made easy* (1754). In his dedication to Hogarth, Kirby wrote: "this Work in a peculiar Manner has a Right to your Patronage and Protection, as it was YOU who first encouraged me upon the Subject."

> and considered it, at the same time that they complain of its obscurity. The attention and application which the reading, and understanding this little book [*Linear Perspective* or *New Principles*] require ... has discouraged the generality of those for whose service it was chiefly designed, from the attempt; so that very few have profited by the best treatise that has been published on the subject. [Highmore, 1763, p. v]

Kirby's and Highmore's opinion, that Taylor had created a very valuable theory of perspective which deserved to be transmitted—in a revised version—to a larger public, became widely shared among an English group of painters, draftsmen, and engravers. In the period from 1761 to 1803 there appeared—besides Kirby's and Highmore's books—three more books whose titles explicitly mention Taylor's theory; these were written by Daniel Fournier (1761), Thomas Malton (1775), and Edward Edwards (1803). Several other

Figure 37. This engraving has the following text: "To Mrs Younge, Daughter of Dr. Brook Taylor. This Plate as a Tribute Due to her Father's Merit, is Dedicated by Her unknown but most Respectfull humble Servant Joshua Kirby." From Kirby's *Dr. Brook Taylor's method of Perspective Made Easy* (1768).

English authors, practitioners as well as scientists, based their works on perspective on Taylor's theory. Among these are William Emerson (1765), John Lodge Cowley (1766), Joseph Priestley (1770), Edward Noble (1771), and Thomas Malton's son James Malton (1800). In their writings most of these men praised Taylor, and some of them did it also in their illustrations as Figure 37 and Figure 38 show.

Apart from Hamilton, it is characteristic of Taylor's successors that they did not add anything essential to his theory; several of them even thought that this would be impossible. Priestley, for instance, claimed that:

> As in all the other branches of mathematical knowledge the progress of this art [perspective] has been slow, but sure; and the English writers (particularly Dr. Brooke Taylor) seem to have carried it to a degree of perfection we can hardly conceive it possible to be exceeded. [Priestley, 1770, pp. v–vi]

Almost a century later we find the same admiration for Taylor's theory expressed in a note which the English mathematician Augustus De Morgan wrote on the history of perspective:

> Nothing could beat the geometry given by Taylor, which wants little additional explanation. The late Peter Nicholson who to sound mathematical knowledge added immense experience of practice says of Taylor's work that "although mere pamphlets they contain all the elementary knowledge necessary on the science of Perspective." [De Morgan, 1861, p. 728]

Figure 38. The text in the picture says "To the Memory of Dr. Brook Taylor in Gratitude for his sublime Principles on PERSPECTIVE." From Thomas Malton's *A Compleat Treatise on Perspective* (1779).

Several decades later an American mathematician, Julian L. Coolidge, also saw Taylor's work as a kind of end product in the development of theoretical perspective:

> ... the capstone of the whole edifice [the theory of perspective] seems to me to have been planned by Brooke Taylor in 1715. [Published 1940, quoted from Coolidge, 1963, p. 108]

Coolidge reached this conclusion although he only knew Taylor's work from Kirby's presentation. About this De Morgan said in his note mentioned above, "Taylor would not have thanked Kirby" (De Morgan, 1861, p. 728); De Morgan's view reflects the attitude of a mathematician who judges the theoretical content first of all, and is less concerned about the process of making a theory accessible to users not that well trained in mathematics. The latter is, however, important; and perhaps Taylor would have appreciated the many efforts to make his theory applicable although it became mathematically less elegant.

The art of teaching perspective seems to be particularly difficult; thus almost all of Taylor's successors introduced their books by writing that they had not yet seen a satisfactory book on perspective. Thus Priestley wrote that although he had been taught perspective, he realized—while making illustrations for his work—that he was not capable of making a draft of an electrical apparatus thrown into perspective, and he found no help in books (Priestley, 1770, p. viii). Malton a few years later expressed himself much stronger:

> ... it is a certain truth; there are indeed a sufficient number of Authors on the Subject [perspective]; and yet perhaps no subject has been worse handled, in general. [Malton, 1776, p. I]

Despite the fact that English writers on perspective continued to criticize the quality of the earlier literature—as Taylor himself had done—the consensus was that perspective ought to be taught according to Taylor's theory, and that Taylor was—in Highmore's words "the inventor of the true universal system of perspective" (Highmore, 1763, Preface). As mentioned on page 3, this idea persisted in some circles until at least the 1880s.

10. The Acknowledgment of Taylor's Theory on the Continent

The recognition of the tremendous impact of Taylor's work on the theory of perspective in England naturally leads to the question whether his ideas also spread to the other side of the channel. The answer is affirmative, but rather than guiding the development on the continent they influenced only a few men. The first sign—that I have come across—of a continental interest in Taylor's work is translations of his *New Principles*. One Italian, Giacopo Stellini, and two Frenchmen, François Jacquier and Antoine Rivoire, engaged themselves independently in making Taylor's text known on the continent. Jacquier, who lived in Rome, made an Italian translation which appeared in 1755 with the result that Stellini gave up a project, already started the previous year, of having his translation published in Venice (Stellini, 1782, p. vii); his translation was first printed in volume three of his posthumous *Opere varie* (1782). Rivoire's French translation was issued in 1757.

The fact that the translations of Taylor's work appeared in the same period as the English practitioners started to teach his theory is remarkable, but there seems to be no connection between the two events. The editor of Stellini's translation, Antonio Evangel, was of the opinion that Stellini started his work on Taylor as early as the 1730s (Stellini, 1782, p. vii). Evangel did not give any other reason for Stellini's work than that it was in agreement with his general intellectual interests. Still it is slightly unexpected that Stellini, who became Professor of Ethics at the University of Padova in 1739 and is best known for his work on ethics and pedagogics, took an interest in such a technical piece of mathematics as *New Principles.* Apparently he saw it as a pedagogical challenge to explain Taylor's theory; thus to his translation he added some clarifying notes and moreover, according to Evangel, he aimed at explaining all Taylor's examples carefully (Stellini, 1782, p. viii). In this enterprise he went no further than the first example, so in 1754 he decided to give up and sent his translation and notes to print.

The man who prevented the completion of Stellini's project, Jacquier, did not give any information about what motivated him to work on Taylor's theory either. However, from the five appendices he added to his translation of Taylor's text it is evident that he was mainly fascinated by the scientific problems of perspective. His appendices contain an examination of the "curve of foreshortening" (p. 54), a study of deformations in curved mirrors leading to a discussion of caustics, and various other investigations in which he applied algebra, calculus and some of Newton's results from *Principia.* As Stellini started his work on Taylor before the English practitioners and as Jacquier had a much more scientific approach to perspective than they did, we may conclude that these two translators worked independently of the development in England.

Whether this is also true for the third translator, the Jesuit Antoine Rivoire, cannot be decided, before more is known about when and in what context he started to translate Taylor. In his preface to *Nouveaux principes de la perspective linéaire* he expressed much concern about practitioners'—particularly architects'—education in perspective, and said that to improve this he had translated Taylor's work. He gave the following reasons for choosing it:

> We do not lack books on this subject [perspective]; and although this part of mathematics has not been pursued as far as the others we do have some good books on perspective. However, some of them belong to an entire course which—to get that small part he needs—an artist only would decide to buy with hesitation. Those which are printed separately can only be acquired for an immense price, either because they are voluminous, or because of the quantity of the plates with which they have been overloaded. This double inconvenience has been avoided in the book whose translation I bring here, without falling into the usual trap of becoming obscure and unintelligible when insisting on being brief.
>
> In this book the author [Taylor] establishes the most general and comprehensive principles, and develops them with clarity and exactness. And even

> though he treats his subject as a great geometer he knows how to make it accessible for those who only have a very feeble knowledge of geometry.[21]

One seldom encounters price as an argument for editing a book on perspective, but it is true that Taylor's *New Principles* offered the most developed theory in a rather small number of pages. If Rivoire really thought that the book could also be understood by practitioners who were almost unacquainted with geometry he was rather optimistic. Not quite keeping to his idea of publishing a brief book on perspective, Rivoire enlarged his edition with translations of Newton's theory of mixing colors and of that part of Patrick Murdoch's *Newtoni genesis curvarum per umbras* ... which treats perspective. Furthermore, he added a section written by Jacques Silvabelle, which he claimed would be helpful for understanding Taylor's appendix on determining projections on curved surfaces—how that should be escapes me.

Although the 1750s brought a Renaissance for Taylor's ideas in England as well as on the continent, the ways in which they were treated were different. In England the writers on perspective attempted to explain Taylor's insights to a public of nonmathematicians, whereas their continental colleagues kept to Taylor's own text. The situation did not change later; thus on the continent the translations were not succeeded by a second generation of books that were mainly based on Taylor's theory. I am not familiar with the entire continental literature on perspective from the second half of the eighteenth century but I dare say that there are only a very few more than the single example of a Taylor inspired book I have come across, the *Traité de perspective linéaire* (1771) by Michel mentioned above (p. 25).

Michel claimed that "the entire art of linear perspective is invented by the celebrated Englishman Doctor Brook Taylor" (translation from Michel, 1771, p. 6). In Taylor's "invention" Michel found the inspiration to use vanishing points and lines frequently and elegantly; the book clearly shows that Michel was very well acquainted with the theory underlying Taylor's constructions, but he did not present this theory, restricting himself to explaining how

---

[21] Translation from Taylor, 1757, in the 1759 edition, pp. v–vi which reads:

> "Ce n'est pas que nous manquions de Livres sur cette matiere [perspective]; & quoique cette partie des Mathématiques n'ait pas été poussée aussi loin que les autres, nous avons pourtant quelques bons ouvrages sur la Perspective: mais quelques-uns sont à la suite d'un cours entier, qu'un Artiste ne se détermineroit à acheter qu'avec peine, pour acquerir cette seule partie dont il a besoin. Ceux qui ont été imprimés séparément ne laissent pas d'être d'un prix excessif, soit par la grosseur du volume, soit par la quantité des planches dont ils sont surchargés. On a évité ce double inconvénient dans le Livre dont je donne ici la traduction, sans tomber dans celui où l'on tombe communément, quand à force de vouloir être court, on se rend obscur & inintelligible.
>
> L'Auteur [Taylor] y établit les principes les plus généraux & les plus étendus, il les y develope avec clarté & précision, & lors même qu'il traite son sujet en grand Géométre, il sçait le mettre à la portée de ceux qui n'auroient que les plus foibles connaissances de la Géométrie."

constructions should be performed. Thus Michel did not contribute to the spread of understanding of Taylor's theory.

Despite the fact that Taylor's theory was presented only in a rather small number of continental books, a knowledge of and an admiration for it were kept alive on the continent. In his *Histoire des mathématiques*, Jean Étienne Montucla emphasized that Taylor had treated perspective in a new manner (Montucla, 1758, p. 638 or 1799, p. 711). Referring to Montucla the encyclopedist De la Chapelle characterized 's Gravesande's *Essai de Perspective* and Taylor's *New Principles* as "the two best works that we have on this subject" (translation from Diderot and D'Alembert, 1780, p. 454). Michel Chasles repeated this opinion saying that:

> 's Gravezande and Taylor are often mentioned, and rightly so, for having treated perspective in a new and learned manner.[22]

As late as 1865 Luigi Cremona—using the anagram Marco Uglieni as a pseudonym—published a paper on Taylor's theory. Cremona presented nine of Taylor's eighteen problems from the first part of *New Principles*, keeping rather close to Taylor's own formulation. Furthermore, he paraphrased Taylor's constructions of the problems and gave new proofs of their correctness. As already mentioned in the introduction, Cremona called his paper *I principii della prospettiva lineare secondo Taylor*, thus to him Taylor's principles seem to have been the nine fundamental problems that he had selected.[23] What they were to Taylor's English successors will be investigated in the next section.

11. Taylor's Method and Principles

We have seen that Taylor called his two books on perspective, respectively, *Linear perspective or a New Method* and *New Principles of Linear Perspective*, and that his successors used the expressions Taylor's method and Taylor's principles; it is now time to discuss what these terms cover. One consults Taylor in vain to find out what he himself conceived of as a new method or as new principles in the theory and practice of perspective. Nor did any of his successors explicitly state what they thought to be his method or principles. It is very likely that some of the authors took over the term "principles" from Taylor's second book although they meant procedures rather than principles. However, the search for what the term "Taylor's principles" alludes to does

---

[22] Translation from Chasles, 1837, p. 347, which reads:

> "S'Gravezande et Taylor sont cités souvent, et à juste titre, comme ayant traité la perspective d'une manière neuve et savante ...."

[23] The nine problems Cremona chose have the following numbers in *New Principles*: I, II, III, V, VII, XI, XIV, XVI, XVII.

not become easier by widening the sense of "principles." If we take all the books inspired by Taylor (mentioned in Section 9) together we find applications of all of Taylor's work; but if we look at them separately we find that the various authors selected different parts of Taylor's theory and that they chose different constructions as the basic ones of making perspectives of plane figures. Some put most emphasis on the "division construction" (presented on p. 14), others on the "visual ray construction" (p. 18), others again developed their own constructions using Taylor's theory, and finally there were some—like Kirby—who preferred the traditional "distance point construction" (p. 13) which Taylor had not treated at all. Kirby, who claimed to present Taylor's method, for instance, even preferred this method. Similarly, the various authors had different approaches for throwing three-dimensional figures into perspective.

Having realized how different the various presentations of Taylor's ideas are we are left with the problem of finding a unifying element in them. In Highmore's *Practice of Perspective* we find a clue. The second part of this book Highmore called "a comparative perspective," and in it he aimed at establishing the superiority of Taylor's approach by comparing his constructions with traditional ones. According to Highmore the use of pre-Taylorian methods corresponds to writing down 278 three hundred times and adding all together when the result of 300 times 278 is required, because:

> the common methods are attended with such tedious operations, such a multitude of unnecessary lines, and in some situations, with such perplexed and intricate schemes, as require more than human patience to execute .... [Highmore, 1763, p. 11]

Highmore's way of avoiding the situation just described was to make full use of all vanishing points—and not only the principal vanishing point and the distance points. In a few examples he also applied the concept of a vanishing line. Thus Highmore gave the impression that "Taylor's principles" consisted of an application of the theory of vanishing points and to some extent the theory of vanishing lines. Later Thomas Malton—to whom I shall shortly return—implicitly gave the same idea, only he put more stress on vanishing lines than Highmore had done. I have not found similar hints as Highmore's and Malton's for unraveling the problem of "Taylor's principles" in the other books belonging to the Taylor tradition. Nevertheless, I think that Taylor's principles and method were generally understood as being the theory and practice connected to the general notion of vanishing points and lines. Moreover, I suspect—without being able to prove it—that this was in accordance with what Taylor himself conceived of as the novelty of his approach to perspective.

If my hypothesis is correct then the entire idea of some particular Taylorian principles is based on a misunderstanding, for—as mentioned in Sections 3 and 4—Taylor did not invent the concept of vanishing points and lines. He was the first to spell out the importance of these concepts, but already at the beginning of the seventeenth century Guidobaldo and Stevin had applied the notion of general vanishing points, and vanishing lines had also occurred before Taylor. My interpretation that Taylor's successors credited him with the creation of a theory which actually existed much earlier is supported by some statements of Thomas Malton. In the preface to his *A Compleat Treatise on Perspective ... on the True Principles of Dr. Brook Taylor* he expressed the general opinion that Taylor had made a special contribution to the theory of perspective:

> ... the Principles, on which he has founded his System are the most simple and perfect that can possibly be conceived. [Malton, 1775, p. i]

Later, however, Malton became interested in the history of perspective, and when in 1783 he published a second part of his *Compleat Treatise* he included a long section on the works by some of his predecessors—among them Guidobaldo's *Perspective libri sex*. Commenting upon Problem III of its Second Book—where Guidobaldo applies the main theorem—Malton noticed: "In this Example may be clearly seen the *true Principles of Perspective*" (Malton, 1783, p. 81, my emphasis). In other parts of Guidobaldo's book Malton also found ideas similar to the ones Taylor had put forward later, but he could still praise Taylor for having extended Guidobaldo's theory. The praising of Taylor as an inventor of special principles became, however, impossible for Malton when he compared 's Gravesande's and Taylor's works, here he was led to a conclusion he did not seem to be too happy to draw:

> It is far from my Intention, nor have I a wish to lessen the Merits of my own countryman, Dr. Taylor; I had much rather attributed to him the sole invention of the new Principles, could it be done with candour; but 'tis my determination to give the praise due to the Author, whereever I find occasion. In this work [*Essai de perspective* by 's Gravesande] it is manifest, that a foundation is laid for those universal Principles .... [Malton, 1783, p. 85]

Malton's understanding of the history of "Taylor's principles" did not seem to have had any influence; the idea that Taylor furnished the theory of perspective with a special foundation survived long after Malton, as we have seen. The origin of this idea can be ascribed, in part, to the fact that England was mathematically slightly isolated in the eighteenth century and, in part, to the fact that in general not much attention was paid to mathematical studies of perspective. Thus, also on the continent, the works by Guidobaldo, Stevin, and 's Gravesande were much less known than the more practical approaches to describing the laws of perspective. The reason that the idea survived is presumably related to the fact that in the nineteenth century, mathematicians—considering perspective as a part first of descriptive and later of projective geometry—lost interest in perspective as a theory *per se* and the details of its history. In the later part of the nineteenth century the history of the discipline was studied by N.G. Poudra, but he paid by far most attention to the French contributions. He spent, for instance, more than fifteen pages on describing Desargues's method of perspective, but less than three on presenting Taylor's ideas and did not attempt to pursue the history of "Taylor's principles." The next writer on the general history of perspective—as seen from a mathematical point of view—Loria, was rather enthusiastic about Taylor's *New Principles*:

> The reading of this short but excellent work surprises the modern reader most pleasantly, because in it, in a word, one finds all the fundamental concepts (except perhaps the "circle of distance") and all the methods of central projection, as one finds them presented in for instance the classical textbook by W. Fiedler.[24]

As this quote indicates, Loria was more interested in tracing the history of ideas that later became important in mathematics than in investigating the context in which these ideas were born. Hence he did not engage himself with the question of "Taylor's principles"; and this question was not taken up by Jones either.

12. Concluding Remarks

The studies carried out in this survey have shown that there are at least two paradoxical aspects of Taylor's role in the history of perspective. First, it is paradoxical that Taylor got so much credit for insights that did not originate with him, whereas little attention was paid to his own innovations. Thus the part of Taylor's theory that was mostly applied—and presumably known as his principles—existed long before him. His real and impressive improvements of the theory of perspective—among which his wider use of vanishing lines, and his contributions to direct constructions, and the theory of inverse problems are especially significant—were, however, not much noticed. They may have inspired Lambert, but that is very uncertain, and otherwise nobody continued Taylor's investigations. In the words of Edward Noble Taylor's "fate has been to be more admired and celebrated than understood" (Noble, 1771, p. vi).

Second, it is paradoxical that precisely Taylor's studies, which are the most incomprehensible of the entire pre-nineteenth-century literature on perspective, evoked response in a circle of practitioners. The practitioners who wanted

[24] Translation from Loria, 1908, pp. 597–598:

> "Die Lektüre dieses kürzen, aber vortrefflichen Werkes bereitet dem modernen Leser eine der angenehmsten Überraschungen, da man, um es ganz kurz zu sagen, darin alle Grundbegriffe (nur etwa den "Distanzkreis" ausgenommen) und alle Methoden der Zentralprojektion vorfindet, wie man sie z.B. in dem klassischen Lehrbuch von W. Fiedler ausgeführt findet."

to include the theory of perspective in their textbooks could have found what they needed much more easily in other books, for instance, 's Gravesande's *Essai de perspective*—which was translated into English. Ironically enough, however, it turned out that the most complicated and elaborate theory of perspective became the one that contributed to bridging the gap between the theory and practice of perspective.

Albeit paradoxically, Taylor figures prominently in the history of perspective. It is particularly remarkable how he used his mathematical skill to find the core of the theoretical problems of perspective and that he found it important to communicate his understandings to the users of perspective. In these respects he is in company with mathematicians like Stevin, 's Gravesande, and Lambert. Lambert is the last in this line because after his time perspective ceased to have a mathematical theory of its own, as it became part of descriptive and projective geometry.

Taylor did not succeed in making the deeper parts of his theory common property, but he did succeed in making the English practitioners aware of the importance of the main theorem for understanding the laws of perspective. Moreover he inspired some to work on complicated problems such as the one of determining the perspective reflection in an oblique mirror (Figures 39 and 40).

Finally, it is worth noticing that Taylor seems to have coined the term linear perspective; this does not occur often—if at all—in pre-Taylorian books but became common after Taylor had published his two books on perspective.

Figure 39. From Highmore's *The Practice of Perspective.*

Figure 40. From Malton's *A Compleat treatise on Perspective.*

Appendix: The Books on Perspective which Taylor Presumably Possessed

A Catalogue of the Libraries of Joseph Hall Esq. late one of the Six Clerks in Chancery and of Brook Taylor... (Catalogue 1732) contains, besides general books on painting and architecture, a series of books in which perspective is treated. I can find no arguments that guarantee that all these books belonged to Taylor, nevertheless, I am fairly convinced that they did. Since it is not without interest to know which books in the field he has consulted I list them below (the full titles are in the Bibliography, pp. 249ff).

Bosse, 1643_1, 1643_2, 1653, and 1667.
Dechales, 1674.
Dubreuil, 1672, in a second printing from 1698.
Guidobaldo, 1600.
Lamy, 1701.
Maignan, 1648.
Marolois, 1614, in the 1638 edition.
Ozanam, 1693, in the 1712 English edition.
Scheiner, 1631.
Tacquet, 1669.

Book Two

Linear Perspective

Brook Taylor

LINEAR PERSPECTIVE:

OR, A

New METHOD

Of Reprefenting juftly all manner of

OBJECTS

As they appear to the EYE

IN ALL

SITUATIONS.

A Work neceffary for PAINTERS, ARCHITECTS, &c. to Judge of, and Regulate Defigns by.

BY

Brook Taylor, LL.D. and R.S.S.

LONDON:

Printed for R. KNAPLOCK at the *Bifhop's-Head* in St. *Paul's* Church Yard. MDCCXV.

ERRATA.

PAG. 2.	Lin. 11.	for	E T	read	E F	
3.	15.		A P		A B	
8.	15.		*c d f*		*e d f*	
	16.		*e d c f*		*c d e f*	
23.	1.		D T		D F	
28.	15. 16. 17. 18. 19. 20.	*every where*	O / T		R / X	*and on the contrary.*
31.	20.		R C		R B	
42.	23.		*in Cir.*		*in the Cir.*	

TO THE

READER.

IN this Treatise I have endeavour'd to render the Art of PERSPECTIVE *more general, and more eaſy, than has yet been done. In order to this, I find it neceſſary to lay aſide the common Terms of Art, which have hitherto been uſed, ſuch as* Horizontal Line, Points of Diſtance, *&c. and to uſe new ones of my own; ſuch as ſeem to be more ſignificant of the Things they expreſs, and more a-*

greeable

To the READER.

greeable to the general Notion *I have formed to my self of this Subject.*

Thus much I thought necessary to say by way of Preface; *because it always needs an Apology to change Terms of Art, or any way to go out of the common Road. But I shall add no more, because the shortness of the Treatise it self makes it needless to trouble the Reader with a more particular Account of it.*

LINEAR

LINEAR PERSPECTIVE:

SECTION I.

Containing an Explanation of thoſe things that are neceſſary to be underſtood in order to the Practice of PERSPECTIVE.

PERSPECTIVE is the Art of drawing on a Plane the Appearances of any Figures, by the Rules of Geometry.

In order to underſtand the Principles of this Art, we muſt conſider, That a Picture painted in its utmoſt degree of Perfection, ought ſo to affect the Eye of the Beholder, that he ſhould not be able to judge, whether what he ſees be only a few Colours laid artificially on a Cloth, or the very Objects there repreſented, ſeen thro' the Frame of the Picture, as thro' a Window. To produce this Effect, it is plain the Light ought to come from the Picture to the Spectator's Eye, in the very ſame manner, as it would do from the Objects themſelves, if they really were where they ſeem to be; that is, every Ray of Light ought to come from any Point of the Picture to the Spectator's Eye, with the ſame Colour, the ſame ſtrength of Light and Shadow, and in the ſame Direction, as it would do from the correſponding Point of the real Object, if it were placed where it is imagined to be. So that (*Fig.* 1.) if E F be a Picture, and *a b c d* be the Repreſentation of any Object on it, and A B C D be the real Object placed where it ſhould ſeem to be to a Spectator's Eye in O; then ought the Figure *a b c d* to ſeem exactly to cover the Figure A B C D, and the Rays A O, B O, C O, *&c.* that go from any Points A, B, C, *&c.* of the original Object to the Spectator's Eye O, ought to cut the Picture in the correſponding Points *a*, *b*, *c*, *&c.* of the Repreſentation. Wherefore, in order to demonſtrate any Propoſition in this Treatiſe, upon this Principle, I always ſuppoſe the real original Object to be placed where it ſhould appear to be.

DEFINITIONS.

DEF. I.

THE Center of the Picture is that Point where a Line from the Spectator's Eye cuts it (or its Plane continued beyond the Frame, if need be) at Right Angles.

If the Plane C D (*Fig.* 2.) be the Picture, and O the Spectator's Eye, then a Perpendicular let fall on the Picture from O, will cut it in its Center P.

DEF.

D E F. II.

The Diſtance of the Picture, or principal Diſtance, is the Diſtance between the Center of the Picture and the Spectator's Eye.

In the ſame Figure P O is the Diſtance of the Picture.

D E F. III.

The Interſection of an Original Line is that Point where it cuts the Picture.

If I K be an Original Line cutting the Picture in C, then is C the Interſection of the Line I K,

D E F. IV.

The Interſection of an Original Plane, is that Line wherein it cuts the Picture.

A P is an Original Plane cutting the Picture in the Line C Q, which therefore is its Interſection. *

D E F. V.

The Vaniſhing Point of an Original Line, is that Point where a Line paſſing thro' the Spectator's Eye, parallel to the Original Line, cuts the Picture.

Such is the Point V, the Line O V being parallel to the Original Line I K.

C O R O L. I.

Hence it is plain, that Original Lines, which are parallel to each other, have the ſame Vaniſhing Point. For one Line paſſing thro' the Spectator's Eye, parallel to them all, produces the Vaniſhing Point of 'em all, by this Definition.

C O R O L. 2.

Thoſe Lines that are parallel to the Picture have no Vaniſhing Points. Becauſe the Lines which ſhould produce the Vaniſhing Points, are in this Caſe alſo parallel to the Picture, and therefore can never cut it.

C O R O L. 3.

The Lines that generate the Vaniſhing Points of two Original Lines, make the ſame Angle at the Spectator's Eye, as the Original Lines do with each other.

D E F. VI.

The Vaniſhing Line of an Original Plane, is that Line wherein the Picture is cut by a Plane paſſing thro' the Spectator's Eye parallel to the Original Plane.

Such is the Line V S, the Plane E F, being parallel to the Original Plane A B.

C O R O L. I.

Hence Original Planes, that are parallel, have the ſame Vaniſhing Line. For one Plane paſſing thro' the Spectator's Eye, parallel to them all, produces that Vaniſhing Line.

C O R O L.

COROL. 2.

All the Vaniſhing Points of Lines in parallel Planes, are in the Vaniſhing Line of thoſe Planes. For the Lines that produce thoſe Vaniſhing Points, (by *Def.* 5.) are all in the Plane that produces that Vaniſhing Line, (by this *Def.*)

COROL. 3.

The Planes which produce the Vaniſhing Lines of two Original Planes, being parallel to the Original Planes, and paſſing both thro' the Spectator's Eye, (by this *Def.*) have their common Interſection paſſing thro' the Spectator's Eye, parallel to the Interſection of the Original Planes, and are inclined to each other in the ſame Angle as the Original Planes are. And hence,

COROL. 4.

The Vaniſhing Point of the common Interſection of two Planes, is the Interſection of the Vaniſhing Lines of thoſe Planes.

COROL. 5.

The Vaniſhing Line, and Interſection of the ſame Original Plane, are parallel to each other. Becauſe they are generated by parallel Planes. (By this *Def.* and *Def.* 4.)

DEF. VII.

The Center of a Vaniſhing Line is that Point where it is cut by a Perpendicular from the Spectator's Eye.

Such is S, O S being perpendicular to the Vaniſhing Line V S.

COROL.

COROL.

A Line drawn from the Center of the Picture, to the Center of a Vaniſhing Line, is perpendicular to that Vaniſhing Line.
* As in the Line P S, to the Vaniſhing Line S V.

DEF. VIII.

The Diſtance of a Vaniſhing Line is the Diſtance between its Center, and the Eye of the Spectator. As O S.

COROL.

The Diſtance of a Vaniſhing Line is the Hypothenuſe of a Right-Angled Triangle, (ſuch as O P S) whoſe Baſe being the Principal Diſtance (O P) its Perpendicular is the Diſtance (P S) between the Center of the Picture, and the Center of that Vaniſhing Line.

DEF. IX.

The Directing Plane is a Plane paſſing thro' the Spectator's Eye parallel to the Picture.

Such is the Plain G H.

DEF. X.

The Directing Point of an Original Line is that Point where it cuts the Directing Plane.

Such is the Point G, to the Original Line I K.

PROP.

Linear Perspective. 7

PROP. I. THEOR. I. *

The Representation of a Line is Part of a Line passing thro' the Intersection and Vanishing Point of the Original Line.

(*Fig.* 2.) For the visual Rays I O, K O, *&c.* which produce the Representations of the Points I, K, *&c.* of the Original Line I K, by their Intersections with the Picture in *i*, *k*, *&c.* are all in a Plane passing thro' the Original Line I K, and the Spectator's Eye O. But the Point C, which is the Intersection of the Original Line I K, is in that Plane, because it is in the Line I K; and the Line O V is in the same Plane, because it is parallel to the Original Line I K, (by *Def.* 5.) Wherefore the Line V C is the Intersection of the Plane I O K with the Picture, and consequently *i k*, which is the Representation of the Line I K, is Part of the Line V C, which passes thro' the Vanishing Point V, and the Intersection C, of the Original Line I K.

COROL. 1.

Hence the Representations of any Number of Lines that are parallel to each other, but not parallel to the Picture, will pass thro' the same Point. For they all pass thro the same Vanishing Point. (By this *Theor.* and by *Cor.* 1. *Def.* 5.) See this express'd in *Fig.* 3.

COROL. 2.

But if the Original Lines are parallel to the Picture, as well as to each other, their Representations will be parallel to each other, and to the Originals. For the Line passing thro' the Spectator's Eye, which in other Cases produces the Vanishing Point by its Intersection with the Picture, is in this Case parallel to it, and therefore produces no Vanishing Point. So that the Representations can never meet each other; nor *

that

that Line paſſing thro' the Spectator's Eye, and conſequently are parallel to each other, to that Line, and to the Originals. See this repreſented in *Fig.* 4. where A B is the Picture, O the Eye of the Spectator, *&c.*

COROL. 3.

And hence it appears that the Repreſentations of plain Figures, parallel to the Picture, are exactly of the ſame Shape as their Originals. For (*Fig.* 5.) the Picture being A B, the Original Figure parallel to it being C D E F, and the Repreſentation being *c d e f*, if you reſolve the Original Figure into Triangles, by means of Diagonal Lines, ſuch as D F, the Repreſentation will be reſolved into Triangles by correſponding Diagonals *d f*. Whence all the Lines in the Repreſentation being parallel to all the Lines in the Original, every Triangle * *e d f* will be like the correſponding Triangle E D F, and con- * ſequently all the Lines of the Figure *e d c f* will be in the ſame Proportion to one another, as the correſponding Lines in the Figure C D E F.

PROP. 2. THEOR. 2.

Any Line in the Repreſentation of a Figure parallel to the Picture, is to its Original Line, as the Principal Diſtance is to the Diſtance between the Spectator's Eye, and the Plane of the Original Figure.

Let OG (*Fig.* 5.) be perpendicular to the Original Plane and to the Picture, cutting them in G, and *g*. The Original Plane being parallel to the Picture, therefore all the viſual Rays O C, O D, O E, *&c.* are cut in the ſame Proportion by the Points *c d e*, *&c.* as OG is by the Point *g*; and *d c* being parallel to D C, the Triangle *d* O *c* is like the Triangle D O C; wherefore *d c* is to D C, as O *d* is to O D, that is as O *g* (which is the principal Diſtance) is to O G.

PROP.

PROP. 3. THEOR. 3.

The Repreſentation of a Line is parallel to a Line paſſing thro' its Directing Point and the Spectator's Eye.

For the Directing Point G, (*Fig.* 2.) and conſequently the Line O G, is in the Plane I O K, which produces the Repreſentation *i k*. (By *Prop.* 1.). Wherefore the Directing Plane being parallel to the Picture, (by *Def.* 9.) the Repreſentation *i k* is parallel to O G.

COROL.

Hence it appears that Original Lines which paſs thro' the ſame Point of the Directing Plane, have parallel Repreſentations. Which alſo is true, when they don't paſs thro' the ſame Point of the Directing Plane, but thro' any Points of the ſame Line (O G) that paſſes thro' the Spectator's Eye.

PROP. 4. THEOR. 4.

The Diſtance between the Vaniſhing Point of a Line and the Repreſentation of any Point in it, is to the Diſtance between the Vaniſhing Point and Interſection, as the Diſtance between the Directing Point and the Interſection, is to the Diſtance between the Directing Point and the Original Point.

(*Fig.* 2.) Every thing remaining as is explained in the Definitions, and *k* being the Repreſentation of K, the Line V C being Parallel to O G, the Triangles O V *k*, and K G O are ſimilar; wherefore V *k* : O G :: O V : G K. But O G = V C, and O V = G C; wherefore V *k* : V C :: G C : G K.

COROL.

If you imagin a Plane to paſs through K parallel to the Picture, this Proportion will be the ſame as in *Prop.* 2.

Section II.

Propoſitions relating to the General Practice of Perspective.

* Prop. 5. Prob. I.

Having given the Center and Diſtance of the Picture,
* *to find the Repreſentation of a Point, whoſe Seat on the Picture, and Diſtance from it are given.*

Fig. 6. S is the Center of the Picture, and C the Seat of the Original Point. Thro' S and C draw at pleaſure two parallel Lines, and make S O equal to the Diſtance of the Picture, and C P equal to the Diſtance of the original Point from it's Seat, and draw O P, which will cut S C in *p*, which is the Point ſought.

When the Picture lies between the Spectator's Eye, and the original Point, *p* will fall between S and C; but when the original Point falls between the Spectator's Eye and the Picture, then will *p* fall beyond C, as is expreſſed by the Letters P1 and *p*1; but when the Spectator's Eye falls between the Picture and the original Point, that Point cannot be ſeen on the Picture, becauſe it is behind the Spectator, though it's Projection is of uſe in ſome Caſes.

DEMON-

DEMONSTRATION.

For you may imagin O to be the Spectator's Eye, and P to be the original Point, in their proper Places, then will O P be a viſual Ray projecting the Point *p* on the Picture.

The Point *p* may be found likewiſe by Calculation, for $Sp : SC :: OS : OS + CP$, or $Sp1 : SC :: OS : OS - P1C$.

Prop. 6. Prob. 2. *

Having given the Center and Diſtance of the Picture, and the Poſition of an Original Line with reſpect to the Picture, to find its Indefinite Repreſentation and Vaniſhing Point.

I. *By the Seat, Interſection, and Inclination of the Original Line.*

Fig. 7. Let S be the Center of the Picture, C the Interſection of the original Line, and C D its Seat on the Picture.

Draw C A, ſo that the Angle A C D may be equal to the Angle the original Line makes with its Seat C D. Draw S V parallel to C D, and perpendicular to it draw S O equal to the Diſtance of the Picture, and draw O V parallel to C A cutting S V in V, and draw C V. Then will V be the Vaniſhing Point, and C V the indefinite Repreſentation of the Original Line.

II. *By the Seats of two Points of the Original Line.* *

Let D and E (*Fig.* 7.) be the Seats of two Points of the Original Line on the Picture. Draw D E, and S V parallel to it. Perpendicular to D E take D A, and E B equal to the Diſtances of the two Original Points from

C 2 their

their Seats; and Perpendicular to S V take SO equal to the Diſtance of the Picture, and draw O V parallel to A B, cutting S V in V. Then will V be the Vaniſhing Point. Draw S E, and find the Repreſentation of B. (By *Prop.* 5.) Then will V *b* be the indefinite Repreſentation. Or find the Repreſentations *a* and *b* of the two Points A and B (by *Prop.* 5) and draw *b a* cutting S V in V. Then will V *b* be the indefinite Repreſentation, and V the Vaniſhing Point.

The Vaniſhing Point V, may likewiſe be found by a Line of Tangents; for the Angle V O S is equal to C A D which is the Compliment of the given Angle A C D; wherefore the Diſtance of the Picture S O being Radius, S V will be the Co-tangent of the Angle A C D.

DEMONSTRATION.

Suppoſe the Triangles S V O and C D A to be turn'd round the Lines S V, and C D, till O co-incides with the Spectator's Eye, and C A co-incides with the Original Line. Then the Planes S O V, and C A D being parallel to each other, and the Angles O V S, and A C D ſtill equal, O V will be parallel to C A. Wherefore V will be the Vaniſhing Point, (by *Def.* 5.) and C V will be the indefinite Repreſentation. (By *Prop.* 1.)

* PROP. 7. PROB. 3.

Having given the Center and Diſtance of the Picture, and the Poſition of an Original Plane, to find its Vaniſhing Line, its Center, and Diſtance.

Find the Vaniſhing Points of two Lines in that Plain, (by *Prop.* 6.) and a Line paſſing through them will be the Vaniſhing Line ſought. (By *Cor.* 2. *Def.* 6.) Or let S be the Center of the Picture, (*Fig.* 8.) and A B the Interſection of the Original Plane. Draw SO parallel to A B, and equal to the Diſtauce

Diſtance of the Picture, and draw S V perpendicular to it, and let C be the Angle of the Inclination of the Original Plane to the Picture. Draw O V cutting S V in V, ſo that the Angle O V S may be equal to C, and draw V D parallel to A B. Then will V D be the Vaniſhing Line ſought, V its Center, and V O its Diſtance.

The Diſtance S V may alſo be found by a Line of Tangents; for S O being Radius, S V is the Co-tangent of C.

DEMONSTRATION.

Turn the Triangle S V O round the Line S V, till its Plane becomes perpendicular to the Picture. Then will the Plane O V D be parallel to the Original Plane, and conſequently will produce the Vaniſhing Line D V. (By *Def.* 6.) And O V being perpendicular to V D, V will be its Center, and V D its Diſtance, (by *Def.* 7. and 8.)

PROP. 8. PROB. 4. *

Having given the Center and Diſtance of the Picture, and the Vaniſhing Line of a Plane, to find the Vaniſhing Point of Lines perpendicular to that Plane.

Fig. 9. S is the Center of the Picture, and V A the Vaniſhing Line given. Draw S O parallel to V A, and equal to the Diſtance of the Picture, and draw S V perpendicular to V A cutting it in V. Then draw V O, and O P perpendicular to it, which will cut V S in the Point ſought P.

Or you may find the Diſtance S P by Calculation, for V S : S O : : S O : O P. *

DEMON-

DEMONSTRATION.

If the Plane VOP be turn'd round the Line VP, till O co-incides with the Spectator's Eye; then will OVA be the Plane producing the Vaniſhing Line VA, and OP being perpendicular to that Plane, will be parallel to the Original Lines that are perpendicular the Original Planes, and therefore will produce their Vaniſhing Point P.

N. B. *When the Vaniſhing Line paſſes thro' the Center of the Picture, the Points S and V will be all one, and the Diſtance SP, or VP will be Infinite, and the Repreſentations of all Lines perpendicular to the Original Plane, will be perpendicular to the Vaniſhing Line, and conſequently parallel to each other, as they are to the Originals.*

* PROP. 9. PROB. 5.

Having given the Center and Diſtance of the Picture, and the Vaniſhing Point of a Line; to find the Vaniſhing Line of Planes perpendicular to that Line.

Fig. 9. S is the Center of the Picture, and P the Vaniſhing Point given. Draw SP, and perpendicular to it take SO equal to the Diſtance of the Picture, and draw OP, and perpendicular to it draw OK cutting PS in V. Then draw VA perpendicular to VP, and that will be the Vaniſhing Line ſought, V its Center, and VO its Diſtance.

This is demonſtrated as the foregoing *Prop.* by bringing O to the Spectator's Eye. And SV may in the ſame manner be found by Calculation.

N. B. *When*

N. B. *When the Point P co-incides with the Center of the Picture, the Diſtance P V, or S V will be Infinite, and there will be no Vaniſhing Line, the Original Planes being parallel to the Picture.*

PROP. 10. PROB. 6. *

Having given the Center and Diſtance of the Picture, and the Vaniſhing Line of a Plane, and the Vaniſhing Point of the Interſection of that Plane, with another Plane perpendicular to it; to find the Vaniſhing Line of that other Plane.

Fig. 9. S is the Center of the Picture, A V the Vaniſhing Line given, and in it (by *Cor.* 4. *Def.* 6.) D is the Vaniſhing Point given. Find P the Vaniſhing Point of Lines perpendicular to the Plane, whoſe Vaniſhing Line is D A. (By *Prop.* 8.) Then draw D P, which is the Vaniſhing Line ſought.

DEMONSTRATION.

For the Plane whoſe Vaniſhing Line is ſought being perpendicular to the other Plain, P will be the Vaniſhing Point of one Line in it. But D is the Vaniſhing Point of another Line in it by the Suppoſition. Wherefore D P is its Vaniſhing Line.

N. B. *When the Vaniſhing Line A D paſſes through the Center of the Picture, the Diſtance V P will be Infinite,* (*by* Prop. 8.) *Wherefore in that Caſe D P will be parallel to V P, and conſequently perpendicular to A D.*

PROP.

* PROP. 11. PROB. 7.

Having given the Center and Distance of the Picture, and the Inclination of two Planes, and the Vanishing Line of one of them, together with the Vanishing Point of their Intersection, to find the Vanishing Line of the other Plane.

Fig. 10. S is the Center of the Picture, AB the given Vanishing Line of one of the Planes, and A the Vanishing Point of its Intersection with the other Plane, the Planes being inclined to each other in the Angle C. Find the Vanishing Line TB of a Plane perpendicular to the Line whose Vanishing Point is A, (by *Prop.* 9.) its Center will be V, where it is cut by AS. Take VO equal to the Distance of the Vanishing Line BT, and thro' the Point B, where it cuts the given Vanishing Line AB, draw BO, and make the Angle BOT equal to C, the Line OT cutting TB in T, and draw AT, which will be the Vanishing Line sought.

DEMONSTRATION.

Turn the Triangle BOT round the Line BT, till O coincides with the Spectator's Eye; then will OA be perpendicular both to OB and to OT, it being perpendicular to the Plane BOT by the Construction, and OA will be parallel to the Intersection of the Original Planes. Wherefore AB being the Vanishing Line of one of them, and the Angle BOT being equal to C, TA will be the Vanishing Line of the other of them. (By *Cor.* 3 and 4. *Def.* 6.)

N. B. *When the Intersection of the Planes is parallel to the Picture, the Point A being at an infinite Distance, S A will be parallel to the Vanishing Line B A, and V will co-incide with S, and B V will be perpendicular to B A; V O will be the Principal Distance, and T A will be parallel to B A.*

PROP.

PROP. 12. PROB. 8. *

Having given the Vaniſhing Line of a Plane, its Center and Diſtance, and the Vaniſhing Point of a Line in that Plane, to find the Vaniſhing Point of Lines that make a given Angle with that Line.

Q V (*Fig.* 11.) is the Vaniſhing Line given, S its Center, and V the Vaniſhing Point given. Perpendicular to Q V draw S O equal to the Diſtance given. Then draw O V and make the Angle V O Q equal to the given Angle, and Q will be the Vaniſhing Point ſought. So that if you draw any two Lines V P, and Q P, the Angle V P Q will repreſent an Angle equal to V O Q.

DEMONSTRATION.

Turn V O Q round the Vaniſhing Line V Q, till O co-incides with the Spectator's Eye. Then will O V and O Q be parallel to the Original Lines whoſe Vaniſhing Points are V and Q, and conſequently the Angle they make with each other is equal to V O Q.

By this Propoſition you may find the Repreſentations of any plain Figures, having one ſide. For the Original Figure may be reſolved into Triangles, whoſe Angles being all given, the Vaniſhing Points of all their ſides will be given by this Propoſition. Whence the Repreſentations of all the ſides will be found, beginning firſt with thoſe that lie next the ſide given.

Here likewiſe you may obſerve, That the Center of the Picture is not concerned in this *Problem.* Wherefore to ſee the Repreſentations of Figures in a Plane, whoſe Vaniſhing Line is V Q, the Spectator's Eye may be placed any where in the Circumference of a Circle deſcribed by O, while the Plane V O Q turns round the Axis V Q.

* PROP. 13. PROB. 9.

Having given the Repreſentation of a Line, and its Vaniſhing Point; to find the Repreſentation of a Point that divides the Original Line in a given Proportion.

Fig. 12. A B is the Repreſentation given, and V the Vaniſhing Point. Draw at pleaſure A *b*, and divide it in the given Proportion in *c*. Draw V O parallel to A *b*, and draw *b* B cutting it in O. Then draw O *c*, which will cut A B in the Point ſought C.

This may likewiſe be done by Calculation; for A C : A B :: A *c* × A V : A *c* × A V + *bc* × B V, or A C 1 : A B :: A *c* 1 × A V : A *c* 1 × A V — *b c* 1 × B V, or A C 2 : A B :: A *c* 2 × A V : *b c* 2 × B V — A *c* 2 × A V. *

DEMONSTRATION.

Suppoſe the Original Line to be in a Plane, whoſe Vaniſhing Line is V O. Then will O A, O B *b*, and O C *c* repreſent parallel Lines. Wherefore the Originals of A B and A *b* are divided in the ſame Proportion by the Originals of the Points C and *c*. But A *b* being parallel to V O, it is divided by *c* in the ſame Proportion as its Original (by *Cor.* 3. *Prop.* 1.) that is, in the Proportion given. Wherefore the Original of A B is alſo divided in the Proportion given.

PROP.

*

PROP. 14. PROB. 10.

Having given the Vaniſhing Point of a Line V (Fig. 13.) *and three Points in its Repreſentation, A, B, C; to find a fourth Point D, ſo that the Part repreſented by C D may be to the Part repreſented by A B, in a given Proportion.*

Draw at pleaſure V O, and A *d* parallel to each other, and thro' any Point O of the Line V O draw V B, V C cutting A *d* in *b*, and *c*. Make *c d* to A *b* in the given Proportion, and draw O *d* cutting A V in D, which will be the Point ſought.

This is demonſtrated as the foregoing *Propoſition.*

The Point D may alſo be found by a Scale and Compaſſes, by the help of this Proportion C D : C V :: *c d* × A B × C V : *c d* × A B × C V + A *b* × A V × B V. Which may eaſily be accommodated to any Poſition of the Points A, B, C, D, only obſerving well the Signs + and —.

PROP. 15. PROB. 11.

Having given the Vaniſhing Line of a Plane, its Center and Diſtance, and the Repreſentations of two Lines in that Plane, from a given Point in one of them to cut off a Line, that ſhall repreſent a Line in a given Proportion to the Line repreſented by the other.

Fig. 14. S D is the given Vaniſhing Line, and V its Center, and A B and C I are the Repreſentations given, whoſe Vaniſhing Points are D and E. It is required to find the Point I, ſo that the Line repreſented by C I, may be to the Line repreſented by A B, as *c* is to *d*. Draw V O perpendicular to

F D, and equal to the Diſtance given, and draw O D and O E, and make O Q : O R : : *c* : *d*, and draw O S parallel to R Q. Draw A C cutting S D in F, and draw F B and D C cutting each other in *p*. Then draw *p* S, which will cut C E in the Point ſought I.

DEMONSTRATION.

Becauſe of the Vaniſhing Points D and F, the Figure A B *p* C repreſents a Parallelogram. Wherefore C *p* repreſents a Line equal to the Line repreſented by A B.

By the manner of placing the Point O, the Plane S O D may be conſidered as the Plane producing the Vaniſhing Line S D, O being the Spectator's Eye. Wherefore O S being parallel to R Q, D, E and S are the Vaniſhing Points of the Sides of a Triangle parallel to R O Q. Wherefore *p* C I repreſents that Triangle, and conſequently the Line repreſented by C I is to the Line repreſented by *p* C, or by A B, as O Q is to O R; that is, as *c* is to *d*.

If the Point F falls out of reach, you may bring it into the bounds of the Picture thus. Inſtead of A B ſuppoſe *a b* to be one of the Lines given, whoſe Vaniſhing Point is D. Draw
* D A, and taking any Point *r* in the Vaniſhing Point F D, as is moſt convenient, draw *r a* and *r b* cutting D A in A and B, and uſe the Line A B, as above, inſtead of *a b*. For becauſe of the Vaniſhing Points *r* and D, A B and *a b* repreſent equal Lines.

PROP. 16. PROB. 12.

Having given the Vaniſhing Line of a Plane, and the Repreſentations of two Lines in it; to find a Point in one of them, ſo that it may repreſent a Line divided in the ſame Proportion, as the Line repreſented by the other.

Fig.

Fig. 15. P *k* is the Vaniſhing Line given, and AC, DE are the two Repreſentations, and it is required to find the Point F, ſo that the Parts repreſented by DF and FE may be in the ſame Proportion as the Parts repreſented by AB and BC. Draw any Line *s v* parallel to the Vaniſhing Line, and by means of any Vaniſhing Point *w* transfer the Points A, B, C, to *s*, *t*, *v*, and *s v* will be divided in the ſame Proportion as the Original of AC (by *Cor.* 3. *Prop.* 1.) and becauſe of *w s*, *w t*, *w v*, repreſenting Parallels. Whence you may find the Point F by *Prop.* 13.

When the Vaniſhing Point *k* of the Line AB falls within reach, you may find the Proportion of *s t*, to *t v*, without drawing any Lines on the Picture. For *s t* : *t v* : : AB × C *k* : BC × A *k*.

By theſe Propoſitions you may find the Perſpective Repreſentation of any Figures propoſed. For you may conſider the whole Deſign as one Figure, the Poſition of whoſe Parts are all given with reſpect to each other. Then having found the Vaniſhing Line of ſome Principal Plane, and the Repreſentation of a Remarkable Line in the Deſign, by *Prop.* 5, 6, 7, you may find the Vaniſhing Lines of all other Plains, by their Poſitions with reſpect to the Plane firſt aſſumed, by *Prop.* 9, 10, 11. And then by the Vaniſhing Points already given, and by the given Proportions of the Parts, you may find the Repreſentations of the Figures propoſed by *Prop.* 12, 13, 14, 15, 16. Or having given the Plan of any Figure, upon any Plane that is conveniently choſen, and the Elevations of the Parts of the Figure above that Plane, having found the Repreſentation of the Plan, by *Prop.* 12, or by that and *Prop.* 13, 14, 15, 16, you may then find the Repreſentations of the Elevations, by the help of *Prop.* 8. And in order to do all this, there is no neceſſity of having an Original Deſign drawn out in its juſt Proportions ; but it is ſufficient to have the Proportions of the Parts expreſs'd in Numbers; the Deſign being any how ſcetch'd out to help the Memory. This I ſhall illuſtrate by a few *Examples.*

EXAM-

* *EXAMPLE* I.

To find the Repreſentations of any Figures on a Plane, having given the Interſection, and Vaniſhing Line, Center and Diſtance, the Original Plane being drawn out in its juſt Proportions.

At the end of *Prop.* 12. I have obſerved that the Shapes of the Repreſentations of Figures on a Plane don't at all depend upon the Angle the Picture makes with that Plane. Wherefore let F E (*Fig.* 16.) be the Interſection given, *f e* the Vaniſhing Line, V its Center, and V O its Diſtance perpendicular to it. Then O being conſider'd as the Spectator's Eye in the Plane X, which generates the Vaniſhing Line, and which is now ſo turn'd as to co-incide with the Plane of the Picture Y, let Z be the Original Plane turned in the ſame manner till it co-incides with the Picture, the Figures A B C D and M being now ſeen on the back-ſide. Then to find the Repreſentation of any Point B, draw B E cutting E F in E, and draw O *e* parallel to it, cutting the Vaniſhing Line in *e*; then E being the Interſection, and *e* the Vaniſhing Point of the Line B E, the Repreſentation of the Point B is found by drawing E *e*, and the viſual Ray O B cutting it in B. The Point C is found without drawing the viſual Ray, by the Interſection of the indefinite Repreſentations E *e*, and F *f* of the Lines C E and C F. And in the ſame manner the Repreſentation *m* of the Figure M, is found by the Interſections and Vaniſhing Points, without drawing any viſual Rays.

Or you may find the Repreſentations of the Figures propoſed by means of the Directing Plane. Thus (*Fig.* 17.) D F C being the Original Plane reverſed, as before, D F G H the Plane of the Picture, D F the Interſection, G H the Vaniſhing Line, V its Center, and V O its Diſtance ſet off perpendicular to it, as before; having drawn *d f* parallel to G H, and as far below O, as D F is below G H, *d* O F will be the Directing Plane brought into the Plane of the Picture, *d f* being its Interſection with the Original Plane. Whence having continued

continued the Sides of the Triangle A B C, till they cut D T *
and *d f* in D, E, F, and *d, e, f*; D, E and F being their Interſections, and *d, e, f,* their Directing Points, drawing D *b* parallel to O *d,* E *c* parallel to O *e,* and F *c* parallel to O *f,* they by their Interſections will give the Repreſentation ſought *a b c,* (by *Prop.* 3.)

EXAMPLE II. *

To find the Repreſentation of any given plane Figure, having given the Repreſentation of one Side of it, and the Vaniſhing Line, Center and Diſtance of the Plane it is in: (*Fig.* 18.) M F is the given Vaniſhing Line, V its Center, and V O its Diſtance ſet off perpendicular to it, and *a b* is the Line given to repreſent the Side AB of the Figure ABCDE. Having reſolved the Original Figure into Triangles, by means of the Diagonals A C, A D, the Vaniſhing Points of the Sides A C, B C, are found by continuing *a b* till it cuts the Vaniſhing Line in its Vaniſhing Point F; and then making the Angle F O H equal to B A C, you have the Vaniſhing Point H of the Side A C; and by making the Angle F O G equal to *p* B C, you have the Vaniſhing Point G of the Side B C: Whence you have the Repreſentation *a b c* of the Triangle A B C. And in the ſame manner by the Side *a c* you will get the Triangle *a c d,* and then by *a d* you get the Triangle *a d e,* and by that means have the Repreſentation *a b c d e* of the whole Figure. This is founded upon *Prop.* 12.

Or you may proceed thus. Continue the Sides D C, D E (*Fig.* 19.) till they cut A B in P and Q. Then having found the Repreſentations *p* and *q* of the Points P and Q (by *Prop.* 13.) you will find the Repreſentation of the Triangle P D Q, as before (by *Prop.* 12.) and then you will have the Points *c* and *e,* by *Prop.* 13.

Or having continued *a b* (*Fig.* 20.) till it cuts the Vaniſhing Line in F, and drawn D E parallel to the Vaniſhing Line H F, cutting *a b* in D, draw D B parallel to O F, and draw the viſual Rays O *a,* O *b,* cutting D B in A and B, and on the

the Side A B make a Figure A C B like the Original Figure, and proceed as in *Exam.* 1. the Line D E being conſidered as the Interſection, and D C E as the Original Plane inverted.

Curve-lined Figures are deſcribed by finding ſeveral Points, and then joyning them neatly by Hand. And this may conveniently be done by putting the Geometrical Deſcriptions of Curves into Perſpective, as in the two following Examples of deſcribing a Circle.

EXAMPLE III.

Having given the Vaniſhing Line of a Plane, its Center and Diſtance; to find the Repreſentation of a Circle, from the given Repreſentation of one Radius.

Fig. 21. C is the Repreſentation of the Center, and C A of the Radius given, D E is the Vaniſhing Line, and O the Spectator's Eye placed as in the foregoing Examples. Draw at Pleaſure C B, and make C B to repreſent a Line equal to that repreſented by C A (by *Prop.* 15.) that is, biſect the Angle E O D by the Line O F, and draw F A, cutting E C in B. And in the ſame manner you may find as many Points B in the Circumference of the Circle, as you pleaſe.

EXAMPLE. IV.

Having given the Vaniſhing Line of a Plane, its Center and Diſtance, thro' three Points given to draw the Repreſentation of a Circle.

Fig. 22. A, B, and C are the three Points given, D *f* the Vaniſhing Line, and O the Spectator's Eye placed as above. Draw C A and C B cutting the Vaniſhing Line in D and E, and draw D O and E O. Draw any Line *d* O, and make the Angle *d* O *e* equal to D O E, or having made an Inſtrument containing the Angle D O E, turn it round the Center O,
* till it comes into the Poſition *d* O *e*, the Leg O D cutting the

the Vaniſhing Line in *d*, and O E, (or O *e*) cutting it in *f*. Thro' *d* and *f* draw *d* A and *f* B cutting each other in *p*, and you will have one Point in the Repreſentation ſought. This Conſtruction is taken from the equality of Angles in the Circle inſiſting upon the ſame Baſe, whence the Vaniſhing Points are found by *Prop.* 13.

EXAMPLE V.

Having given the ſame things as in either of the two foregoing Examples, from a given Point of the Picture to draw a Line that ſhall touch the Repreſentation of the Circle.

Fig. 23. D E is the Vaniſhing Line, O the Eye of the Spectator, C the Repreſentation of the Center, and C A the Repreſentation of the Radius, and P the Point given, from whence a Tangent is to be drawn. Draw P C cutting the Vaniſhing Line in E, and (by *Exam.* 3.) find the Radius C B. Any where apart draw *b p*, and make *b c* : *c p* : : B C × P E : C P × B E; that is, make the line *b c p* like the Original of B C P. With the Center *c* and Radius *c b* make a Circle, and draw the Tangent *p* T, and then find the Repreſentation P *t* of it by *Exam.* 2. And you may proceed in the ſame manner by finding the Original Figure, when three Points are given.

If inſtead of from the Point P, it had been required to draw a Tangent parallel to C E, the Point P being ſuppoſed infinitely diſtant, P E would be equal to P C, and conſequently it would be *b c* : *c p* : : B C : B E. But in this Caſe, as alſo always when P falls beyond E, you muſt take *c p* backwards, *p* falling on the contrary ſide of the Point *c*, to what it does in the preſent Scheme. But I ſhall leave the Reader to conſider all the variety of Caſes that may happen from the different Poſition of the Points.

The Uſe of this Example is to find the apparent part of the Baſe of a Cone or Cylinder.

EXAMPLE VI.

Fig. 24. In this Figure having given the Center of the Picture S, its Diſtance equal to the Line L, and the Line A B to repreſent one ſide of the firſt Step, and K I the Vaniſhing Line of the Plane of the Horizon; the Repreſentation of the Baſe A B C D of the firſt Step, is found by the Vaniſhing Points I, K, L, which are found by *Prop.* 12. Then by *Prop.* 8. having found the Vaniſhing Point M of the Perpendiculars A E, F C, *&c.* the Vaniſhing Line of the Face A C E F being M K, the Vaniſhing Point N of the Diagonal A F is found by *Prop.* 12. Whence all the reſt of the Lines are found, as appears ſufficiently in the Figure.

* *EXAMPLE* VII.

Fig. 25. In this Figure the Repreſentation H I K Q of a regular Tetraedrum is found as follows; Having given the Side H I, the Center of the Picture S, its Diſtance equal to the Line L, and G F the Vaniſhing Line of the Face H I K. Having continued the given Side H I till it cuts the Vaniſhing Line G F, in its own Vaniſhing Point F, the Vaniſhing Points G and M of the Sides I K and H K are found by *Prop.* 12. Then having made an Equilateral Triangle A B C, and drawn A D perpendicular to B C, and made the Iſoſceles Triangle A E C, whoſe Sides A E, and C E are equal to A D; the Angle A E C being equal to the Inclination of two Faces of the Figure propoſed, the Vaniſhing Line F N of the Face H I Q is found by *Prop.* 11. Whence having the Vaniſhing Points P and N of the Sides H Q and I Q, you have the whole Repreſentation ſought.

The Reader may exerciſe himſelf in drawing the Repreſentations of the five regular Solids, in order to which their Plans and Profils may be found in the following manner.

Fig. 26. Having made an Equilateral Triangle A B C, and found its Center G, you have the Plan A B C G of a regular Tetraedron.

Tetraedron. Then making the Triangle B C E, whoſe Sides B E, and C E are equal to the Perpendicular B D, and drawing C F perpendicular to B E, you have C F equal to the height of the Vertex above the Center G.

It would be ſuperfluous to ſhew how to make the Plan and elevation of a Cube.

Fig. 27. To find the Plan and elevation of a regular Octaedron, make the regular Hexagon A E B F C D, and having drawn the Lines, as appears in the Figure, you have the Plan of the Figure, on the Plane of one Face A B C. Then having drawn B D cutting A C in G, and D H perpendicular to it, with the Center G and Radius G B make a Circle cutting DH in H, and D H will be the Diſtance between two oppoſite Faces.

Fig. 28. To find the Plan of a regular Dodecaedron, make a regular Decagon A B C D E F G H I K, whoſe Center is S. Then draw a Diagonal A D, leaving out two Angles B and C, and draw S B cutting it in M, and make two regular Pentagons M N O P L, *m n o p l* in ſub-contrary Poſitions, having the ſame Center S, and draw the Lines N D, O F, P H, L K, *o* A, *p* C, *&c.* and you will have the Plan of a Dodecaedron on one of its Faces M N O P L. *

Draw *a v*, and in it make *a b* and *b v* equal to the Perpendicular L R let fall from one Angle of the Pentagon to the oppoſite Side, and make the Iſoſceles Triangle *c b v*, with the Baſe *c v* equal to B D, or, which is all one, to a Diagonal L N of the Pentagon. In *v c* take *c e* equal to L M, and make *a d*, *d f*, *e f* parallel and equal to their oppoſite Lines *c e*, *c b*, *b a*, and let *s* be the Center of the Figure *a b c e f d*. Then biſect *c v* in *g*, and draw *s g* cutting *b c* in *h*, and make *n m* perpendicular to *a d*, *s m* and *s n* being each equal to half L M, and make *b i*, *f k*, *f l* each equal to *b h*, and having drawn the Lines as appears in the Figure, you will have the Profil of the ſame Figure; *d*, *n* and *h* being the height of the Angles perpendicular over K, B and H, D and F, and *k*, *m*, and *c* being the Elevations of the Angles perpendicular over A and I, C and G, and E, and *f l e* being the Elevations of the upper Face, which is perpendicular over *n o p l m.*

* *Fig.* 29. To find the Plan and Profil of a regular Icoſaedron, make a Right-Angle A B C, B C being the double of A B, and with the Center A and Radius A C make a Circle cutting A B in D. Then make a regular Hexagon E F G H I K, whoſe Center is S, and having drawn the Lines S E, S F, S G, *&c.* make the Diſtances S L, S M, *&c.* to S F, as B D is to B C, that is, as the Side of a Pentagon is to the Diagonal, and draw all the Lines as appears in the Figure, and you will have the Plan of a Icoſaedron on one of its Faces L M N.

To find the Profil of the ſame Figure make the Iſoſceles Tri-
* angle O P Q, whoſe Baſe O Q is equal to the Difference between L M and E G, and the Sides O P, Q P are equal to a Perpendicular let fall from L to the oppoſite Side M N of the Triangle L M N. In Q P make P R = P O, and in Q O make O T = L M, and make R X, X V, V T parallel and equal to their oppoſite Lines O T, O P, P R, and draw the Lines as appears in the Figure, and you will have the Profil ſought, Z being the Elevation of the Points over F and K, Y that over G and I, R that over H, T that over E, V that over *m* and *n*, and X that over *l*; and obſerve that the Lines Z R, and T Y are parallel to O P, or V X.

EXAMPLE VIII.

Fig. 30. Having given the Center of the Picture S, its Diſtance S D, and the Repreſentation C A of one Radius of a Sphere, whoſe Vaniſhing Point is V, the Repreſentation of the Center being C, the Repreſentation of the Sphere is found as follows. Draw S R perpendicular to A V, and in it take R O equal to the Hypothenuſe of a Right-angled Triangle, whoſe Baſe is S R, and perpendicular is S D, that is, make R O equal to the Diſtance of the Spectator's Eye from the Point R. Draw A O, V O, C O, and draw any Line parallel to V O, cutting

cutting A O, and C O in *a* and *c*, and with the Center *c*, and Radius *c a* deſcribe a Circle, to which draw a Tangent O *t* cutting C A in T. Then find the Vaniſhing Line P Q of a Plane perpendicular to the Line, whoſe Vaniſhing Point is C (by *Prop.* 9.) and find the Repreſentation of a Circle in that Plane, the Repreſentation of one Radius being C T, and the Center C, (by *Exam.* 3.) and that will be the Contour of the Sphere.

The Demonſtration of this Conſtruction is eaſy to one that conſiders, That the viſual Rays which touch a Sphere make a Cone, whoſe Baſe is a leſſer Circle of the Sphere, whoſe Repreſentation is the Contour of the Repreſentation of the Sphere.

In the Practice of *Perſpective* it is often Neceſſary to draw Lines towards a Vaniſhing Point, or ſome other Point that is out of reach, in which caſe the following Propoſitions are uſeful.

PROP. 17. PROB. 13.

From a given Point to draw a Line that ſhall tend to an inexceſſible Point of a given Line, having the Diſtance of the inacceſſible Point from a given Point of that Line.

Fig. 31. Q is the inacceſſible Point in the Line P Q, the Diſtance B Q being given, and A is the Point from whence a Line is to be drawn tending towards Q.

Draw at pleaſure A P, cutting P Q in P, and taking the Meaſure of P B by a Scale of equal Parts, you will have the length of P Q. Take any Point *q* in the Line B Q, and make P *a* : P *q* : : P A : P Q. Draw *a q*, and A Q will be parallel to it.

PROP.

PROP. 18. PROB. 14.

Having given two Lines that tend to an inaccessible Point, thro' a given Point to draw a Line that shall tend to the same Point.

Fig. 32. A B and C D are the given Lines tending to the inaccessible Point, and P is the Point thro' which it is required to draw a Line P *p* tending to the same Point.

Draw at pleasure P A (*n* : 1) and P B parallel to it, cutting A B and C D in C, A, B, and D, and make B *p* : B D :: A P : A C, and draw P *p*, which will be the Line sought.

Otherwise, without the Compasses.

n : 2, and *n* : 3. Thro' P draw at pleasure two Lines cutting A B and C D in A and C, B and D. Then draw A D and B C meeting in Q. Thro' Q draw at pleasure two Lines cutting A B and C D in *b* and *c*, *a* and *d*, and draw *a c* and *b d* meeting in *p*, and draw P *p*, which will be the Line sought.

SECTION III.

Of finding the Shadows of given Figures.

In this Place I consider Shadows only as the Projections of given Figures on given Surfaces, by the means of given Luminous Points. For I consider the Luminous Body as a Point, to avoid the Difficulties that would attend the Description of Shadows, if the Magnitude of the Luminous Body were taken into Consideration, it being sufficient for Practice to regard only the Center of the Luminous Body; and having found the Contour of the Shadow in that Case, the Penumbra may be drawn

drawn by a good Judgment founded on much Observation; it being difficult to bring every thing to exact Mathematical Constructions, at least so as to be most convenient for Practice. But the Artist that has made himself well acquainted with the Principles of this Art, will easily find ways to satisfy himself in those cases that require more Exactness.

PROP. 19. PROB. 15.

Having given the Center and Distance of the Picture, and the Vanishing Line of a Plane, and the Representation of a Luminous Point, and of its Seat on that Plane; to find the Representation of the Shadow of a Point, whose Representation and Seat on the same Plane are also given; and to find the Vanishing Point of the Ray of Light.

Fig. 33. E T is the given Vanishing Line, S the Representation of the Luminous Point, R the Representation of its Seat, A the Representation of the Point whose Shadow is to be found, B the Representation of its Seat, and P the Vanishing Point of Lines perpendicular to the given Plane, (found by *Prop.* 8) Draw R C cutting the Vanishing Line in V, and draw V P; then draw S A cutting R V in C, and P V in D. Then will C be the Shadow sought, and D the Vanishing Point of the Ray of Light. *

DEMONSTRATION.

For the Originals of R, B, A, and S are in a plain Perpendicular to the Plane, on which the Shadow is to be cast; wherefore C is the Shadow, it being the Intersection of the Ray S A with the Plane R B. Besides, V is the Vanishing Point of R B. Wherefore P V is the Vanishing Line of R S A B, and consequently D is the Vanishing Point of S A.

N. B. *When the Original Luminous Point is behind the Spectator, ſo that it cannot have any real Repreſentation on the Picture, its imaginary Repreſentation (which is as it were the Shadow of the Spectator's Eye on the Picture,) muſt be on the contrary Side of the Plane to the Point, whoſe Shadow is ſought, that is, if B be between P and A, S on the contrary muſt be between* P *and* R *(as in* n : 2.) *And when the Luminous Point is the Sun, or any other Light ſuppoſed to be at an infinite Diſtance, the Points R and V will co-incide, and D will be the ſame as S, as I have ſhewn in* n : 3, *and* n : 4. *In* n : 3, *the Sun appears in the Picture: But in* n : 4, *it is behind the Spectator, and its imaginary Repreſentation is S.*

Prop. 20. Prob. 16.

Having given the Center and Diſtance of the Picture, and the Repreſentation of a Line, and of the Seat of one Point of it on a Plane whoſe Vaniſhing Line is given, and alſo the Vaniſhing Point of that Line; to find the Repreſentation of the whole Seat of that Line on that Plane.

Fig. 34. V E is the given Vaniſhing Line of the Original Plane, P is the Vaniſhing Point of Lines perpendicular to it, (found by *Prop.* 8.) A C is the given Repreſentation of a Line, D its Vaniſhing Point, and B the Repreſentation of the Seat of one Point of it, whoſe Repreſentation is A. Draw P D
* cutting V E in E, and V B cutting A D in C. Then will V B be the Repreſentation of the whole Seat of the Line A D.

DEMONSTRATION.

For the Original of A B C is a Triangle whoſe Vaniſhing Line is P D. Wherefore V muſt neceſſarily be the Vaniſhing Point of B C, *&c.*

When

When you have got the Deſcription of any Figure by the foregoing *Propoſitions*, you generally have the Vaniſhing Points of all the Lines of it. So that having the Seat of one Point of the Figure on the Plane you intend to caſt the Shadow on, you may find the Seats of all the other Points of the Figure by this *Propoſition*, and their Shadows by the foregoing.

PROP. 21. PROB. 17.

Having given the Center and Diſtance of the Picture, and the Vaniſhing Lines of two Planes, and the Repreſentation of their common Interſection, and the Repreſentation of a Line, and of its Seat on one Plane; to find the Repreſentation of the Seat of that Line on the other Plane.

Fig. 35. A B and C B are the Vaniſhing Lines of the two Planes, B D the Repreſentation of their common Interſection, F E the given Repreſentation of a Line, and A E the Repreſentation of its Seat on the Plane, whoſe Vaniſhing Line is A B. By *Prop.* 8. find the Vaniſhing Points G and H of Lines perpendicular to the Planes, whoſe Vaniſhing Lines are A B and C B. Draw A G cutting F E in F, and B C in C, and draw H F cutting B C in I, and draw C D cutting E F in K. Then draw I K, which will be the Line ſought. *

DEMONSTRATION.

F, A, and C are the Vaniſhing Points of the Lines E F, E A, and K C, whoſe Originals are all in the ſame Plane, becauſe of the Vaniſhing Line G C, and becauſe they meet in E and D. But becauſe of the Point D being in the Repreſentation of the Interſection of the two Planes, and becauſe of the Vaniſhing Line B C, the Original of D K, and conſequently of the Point K, is in the Plane, whoſe Vaniſhing Line is B C. But becauſe of the Vaniſhing Line H F, I is the

F Vaniſhing

Vaniſhing Point of the Seat of E F on the Plane, whoſe Vaniſhing Line is B C; wherefore K being the Repreſentation of the Interſection of that Line with that Plane, K I will be the Repreſentation of its Seat.

By this *Propoſition* you may carry a Shadow from one Plane to another. For E F being the Ray of Light, having its Poſition with reſpect to the Plane, whoſe Vaniſhing Line is A B, you find the Shadow K on the Plane, whoſe Vaniſhing Line is B C. So that by theſe three laſt *Propoſitions* you may find the Shadows of any Figures.

In ſome Caſes perhaps it may be moſt convenient to find the Shadows by putting the foregoing Rules of *Perſpective* into *Perſpective*. For Example (*Fig.* 36.) G, E, and F, being the Vaniſhing Points of the Sides of the Triangle ABC, whoſe Shadow is to be caſt on a Plane, whoſe Vaniſhing Line is T V, and Interſection with the Plane of the Triangle A B C is D I; to find the Shadow caſt by the Luminous Point repreſented by S. I ſeek the Repreſentation D*g* of the Interſection of a Plane paſſing through the Luminous Point parallel to the Plane of the Original Triangle, with the other Plane: Then drawing S G, S E, and S F cutting D*g* in *g*, *e* and *f*, and continuing the Sides of the Triangle, till they cut D I in H, I, and K; I conſider S as if it were a Spectator's Eye, and uſe *g*, *e* and *f* as Vaniſhing Points, and K, H, and I as Interſections of the Lines B C, B A, A C, and ſo find the Shadow *a b c*.

SECT.

SECTION IV.

Of finding the Repreſentations of the Reflections of Figures on poliſh'd Planes.

It is well known that the Reflections of Figures on a poliſh'd Plane, as in a Looking-Glaſs, or on the Surface of ſtanding Water, appear to be juſt as much on one ſide of the Plane, as the real Objects are on the other ſide. So that to find the reflected Appearance of any Point, you muſt draw a Perpendicular to the reflecting Plane from the real Point, and in it take a Point at the ſame Diſtance on the contrary ſide of the Plane. For Inſtance, To find the apparent Place of the Reflection of the Point P (*Fig.* 37.) on the Plane AB draw the Perpendicular P B, and on the contrary ſide of the Plane make B p = B P, and p will be the apparent Place of the Point P, ſeen by Reflection. And from hence it follows that the Appearance of any Figure ſeen by Reflection, is exactly of the ſame Shape, and of the ſame Bigneſs as the real Figure, but in an inverted Poſition.

Upon theſe Principles depend the following *Propoſitions*, by which you may eaſily find the *Perſpective* Repreſentations of the Appearances of any Figures ſeen by Reflexion from given Planes.

PROP. 22. PROB. 18.

Having given the Center and Diſtance of the Picture, and the Vaniſhing Line of a reflecting Plane, and the Repreſentation of the Seat of a Point on that Plane; to find the Repreſentation of the Reflection of that Point.

Fig. 38. AB is the Vaniſhing Line given, P the Repreſentation of the real Point, Q the Repreſentation of its Seat on the reflecting Plane, and V the Vaniſhing Point of Lines perpendicular to that Plane. Make Q*p* to repreſent a Line equal to that repreſented by QP (by *Prop.* 13.) and *p* will be the Point ſought.

This is evident by the Introduction to this *Section.*

PROP. 23. PROB. 19.

Having given the Center and Diſtance of the Picture, and the Vaniſhing Line of a reflecting Plane, and the Vaniſhing Point of a Line, together with the Repreſentation of its Seat on the reflecting Plane; to find the Repreſentation of the Reflection of that Line, and its Vaniſhing Point.

Fig. 39. A B is the Vaniſhing Line of the Reflecting Plane, V the Vaniſhing Point of Lines perpendicular to it, P I the Repreſentation of the real Line, C its Vaniſhing Point, and B I the Repreſentation of its Seat on the Reflecting Plane. With the Vaniſhing Point V make B *c*, and B C to repreſent equal Lines, (by *Prop.* 13.) then will *c* be the Vaniſhing Point of the Reflection; and drawing *c*I, that will be its Repreſentation.

DEMON-

DEMONSTRATION.

For if you ſeek the Reflection of the Line by the Reflections of two Points of it, by the foregoing *Propoſition*, then drawing any Line V P, the Parts P Q and Q *p* will repreſent equal Lines; otherwiſe *p* can't repreſent the Reflection of P. Wherefore when Q co-incides with B, B C and B *c* will alſo repreſent equal Lines. But in that Caſe C, B, and *c*, are the Vaniſhing Points of the real Line P I, its Seat B I, and its Reflection *p* I. Wherefore the Vaniſhing Point *c*, and the Reflection *c* I are rightly determined by the foregoing Conſtruction.

PROP. 24. PROB. 20.

Having given the Center and Diſtance of the Picture, and the Vaniſhing Line of a reflecting Plane; to find the Vaniſhing Line of the Reflection of a Plane, whoſe Vaniſhing Line is alſo given.

Fig. 39. A B is the Vaniſhing Line of the Reflecting Plane, V the Vaniſhing Point of Lines perpendicular to it, and A C the Vaniſhing Line of the real Plane. Draw V C cutting A C and A B in C and B, and find the Point *c*, as in the foregoing *Propoſition*, and draw A *c*, which will be the Vaniſhing Line ſought.

DEMONSTRATION.

For by the foregoing *Propoſition* *c* is the Vaniſhing Point of the Reflection of a Line, whoſe Vaniſhing Point is C. But the Vaniſhing Line A C is the Place of all the Points C; and therefore the Vaniſhing Line A *c* is the Place of all the Points *c*, and conſequently is rightly determined by finding one of them, and drawing A *c*.

By

By theſe three *Propoſitions* having got the Vaniſhing Line of the Reflection of one Face of a given Figure, and the Repreſentation of the Reflection of one ſide of that Face, you may find the whole Reflection as you do the real Figure, by the foregoing *Propoſitions*, only obſerving to make the Reflections in a contrary Poſition to the real Objects.

SECTION V.

Of the Inverſe Practice of PERSPECTIVE, and of the manner of Examining Pictures already drawn.

PROP. 25. PROB. 21.

Having given the Repreſentation of a Line divided into two Parts in a given Proportion; to find its Vaniſhing Point.

Fig. 13. Suppoſe the Part repreſented by A B, to be to that repreſented by B C, as *a* to *b*. Draw at pleaſure A *d*, and taking A *b* at pleaſure, make A *b* : *b c* : : *a* : *b*, and draw *b* B and *c* C meeting in O, and draw O V parallel to A *d*, cutting A B in V, which is the Vaniſhing Point ſought.

This is the reverſe of *Prop.* 13.

You may find V by Arithmetick, making C V = $\frac{a \times AC \times BC}{b \times AB - a \times BC}$ *

PROP.

PROP. 26. PROB. 22.

Having given the Repreſentation of a given kind of Triangle, and its Vaniſhing Line; to find its Center and Diſtance.

Fig. 40. A B C is the given Repreſentation of a Triangle, and its Vaniſhing Line is F E, and conſequently the Vaniſhing Points of its Sides are F, D, and E.

Biſect FD and DE in G and H, and raiſe the Perpendiculars G I, H K, and make the Angle G I D equal to the Angle that ought to be repreſented by F B D, and H K E equal to the Angle that ought to be repreſented by D A E, and with the Centers I, and K, and Radius's I D and K D, make two Circles cutting each other in O. Then draw O P cutting F E at right Angles in P, and P will be the Center, and P O the Diſtance of the Vaniſhing Line F E.

DEMONSTRATION.

It is a known Property of the Circle, that the Angles F O D and D O E, are equal to G I D, and H K E; wherefore the Angles F B D, or A B C, and B A C, repreſent Angles equal to G I D, and H K E, by *Prop.* 12.

PROP.

PROP. 27. PROB. 23.

Having given the Repreſentation of a given kind of Parallelogram; to find its Vaniſhing Line, Center and Diſtance.

Fig. 41. A B C D is the given Repreſentation of a Parallelogram. Continue the oppoſite Sides, till they meet in E and F, then will E F be the Vaniſhing Line. Draw the Diagonal A C, and by means of the Triangle A B C find the Center and Diſtance by the foregoing *Propoſition.*

DEMONSTRATION.

For the Original Figure being a Parallelogram, the oppoſite Sides are parallel; wherefore their Repreſentations muſt meet in their Vaniſhing Points, and conſequently E F will be the Vaniſhing Line.

PROP. 28. PROB. 24.

Having given the Repreſentation A B C D (Fig. 42.) *of a given kind of Trapezium, to find its Vaniſhing Line, Center and Diſtance.*

Draw the Diagonals A C and D B meeting in E, and by the given Proportion of the Originals of A E, E C, B E, E D, find the Vaniſhing Points F and G of the Lines A C and D B (by *Prop.* 25.) and you will have the Vaniſhing Line G F. Then by the given Species of the Original of the Triangle A B E, find the Center and Diſtance (by *Prop.* 26.)

PROP.

PROP. 29. PROB. 25.

Having given the Repreſentation of a right-angled Parallelepipedon; to find the Center and Diſtance of the Picture, and the Species of the Original Figure.

Fig. 43. ABCDEFG is the Repreſentation given. Continue the Sides of the Figure till they meet in their Vaniſhing Points H, I, K, and you will have the Vaniſhing Lines H I, I K, K H of all the Faces of the Figure. Then from H and I draw Perpendiculars to I K and H K, meeting in S, which will be the Center of the Picture, L and M being the Centers of the Vaniſhing Lines I K and H K. Upon the Diameter I K make a Circle cutting H L in O. Then will L O be the Diſtance of the Vaniſhing Line I K, Whence the Diſtance of the Picture is eaſily found. Draw the Diagonal G F cutting I K in its Vaniſhing Point N, and draw O N. Then will N O K be the Angle repreſented by A G F, and N O I will be the Angle repreſented by F G E; whence you have the Species of the Face A F E G. And in the ſame manner you may find the Species of the other Faces.

DEMONSTRATION.

The oppoſite Sides of the Figure being to repreſent Parallels, the Points where they meet muſt be their Vaniſhing Points, and conſequently the Lines H I, I K, K H will be their Vaniſhing Lines. But becauſe of all the Angles being Right-angles, H and I will be the Vaniſhing Points of Lines perpendicular to the Planes whoſe Vaniſhing Lines are I K, and H K, and conſequently L and M will be the Centers of thoſe Vaniſhing Lines, and S the Center of the Picture, as is ſufficiently evident by *Prop.* 8. But becauſe of the Circle, the Angle I O K is a Right-angle. Wherefore the Diſtance L O is rightly found to make E G A repreſent a Right-angle; whence the Angles N O K, and N O I are found, which are repreſented by F G A, and F G E, by *Prop.* 12.

When the Sides A B, G C, E D are parallel to each other, (*n* : 2.) the Point H will be infinitely Diſtant, and the Vaniſhing Line I K will paſs thro' the Center of the Picture, the Points L and S in this caſe co-inciding. But becauſe the Point L in this Caſe is left undetermined, the Species of the Figure is alſo undetermined. So that the Point O may be taken any where in the Semicircle I O K. Whence if the Semicircle I O K be placed perpendicular to the Picture, a Spectator's Eye placed any where in that Circumference, will take the Figure A D for a Right-angled Parallelepipedon; tho' the Species of it will be different, according to the different Places of the Spectator's Eye. But this is, provided the Lines A B, G C, *&c.* be perpendicular to the Vaniſhing Line K I; otherwiſe the Figure can't repreſent a Right angled Parallelepipedon, but one inclined on a Right-angled Baſe.

I leave it to the Artiſts to conſider whether this Obſervation may not be of Uſe in Painting the Scenes of a Theater. For hence it appears, that if I and K were the two Points, that in the common Books of Perſpective are call'd the Points of Diſtance, and the Buildings in the Scene were ſo drawn, that their Sides ſhould run to thoſe Points, (as in the preſent Figure,) they would always appear Right-angled to the Spectators in in the front Seats of the Boxes, their Eyes being all in Circumference of the Semicircle I O K; for tho' the Proportions of their Sides were ſomething alter'd, that would be no great Inconvenience.

F I N I S.

Plate. 1.

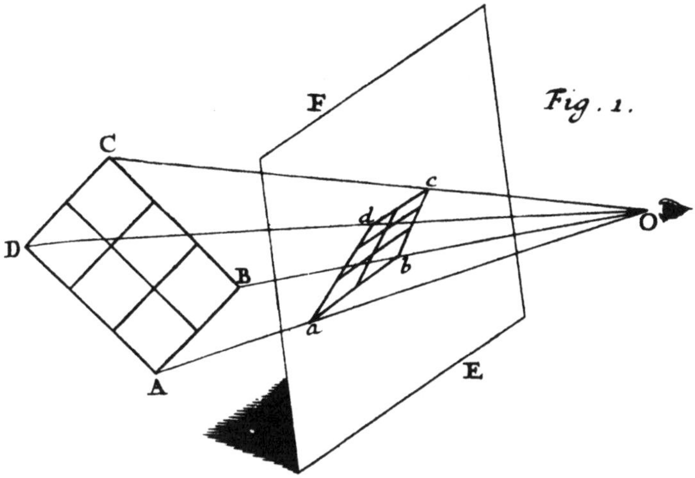

Fig. 1.

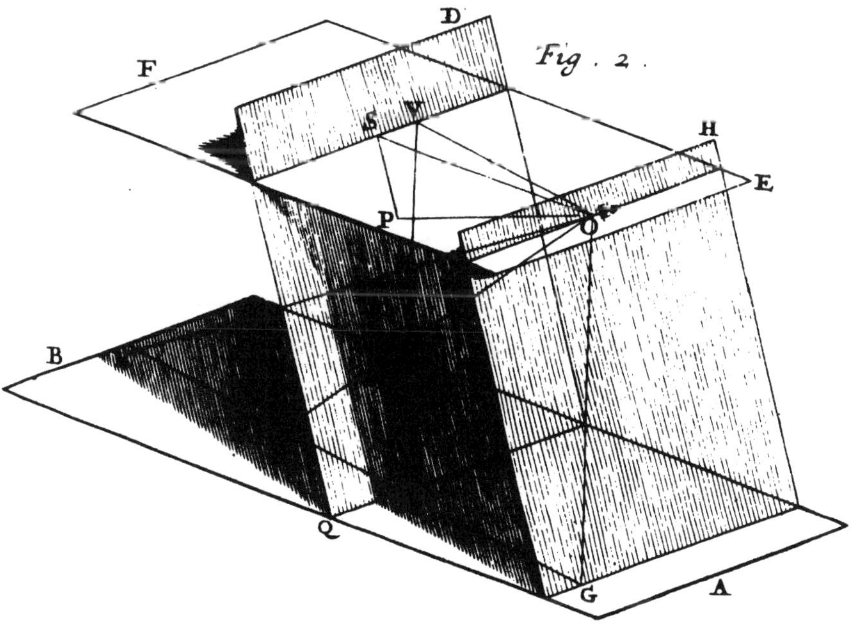

Fig. 2.

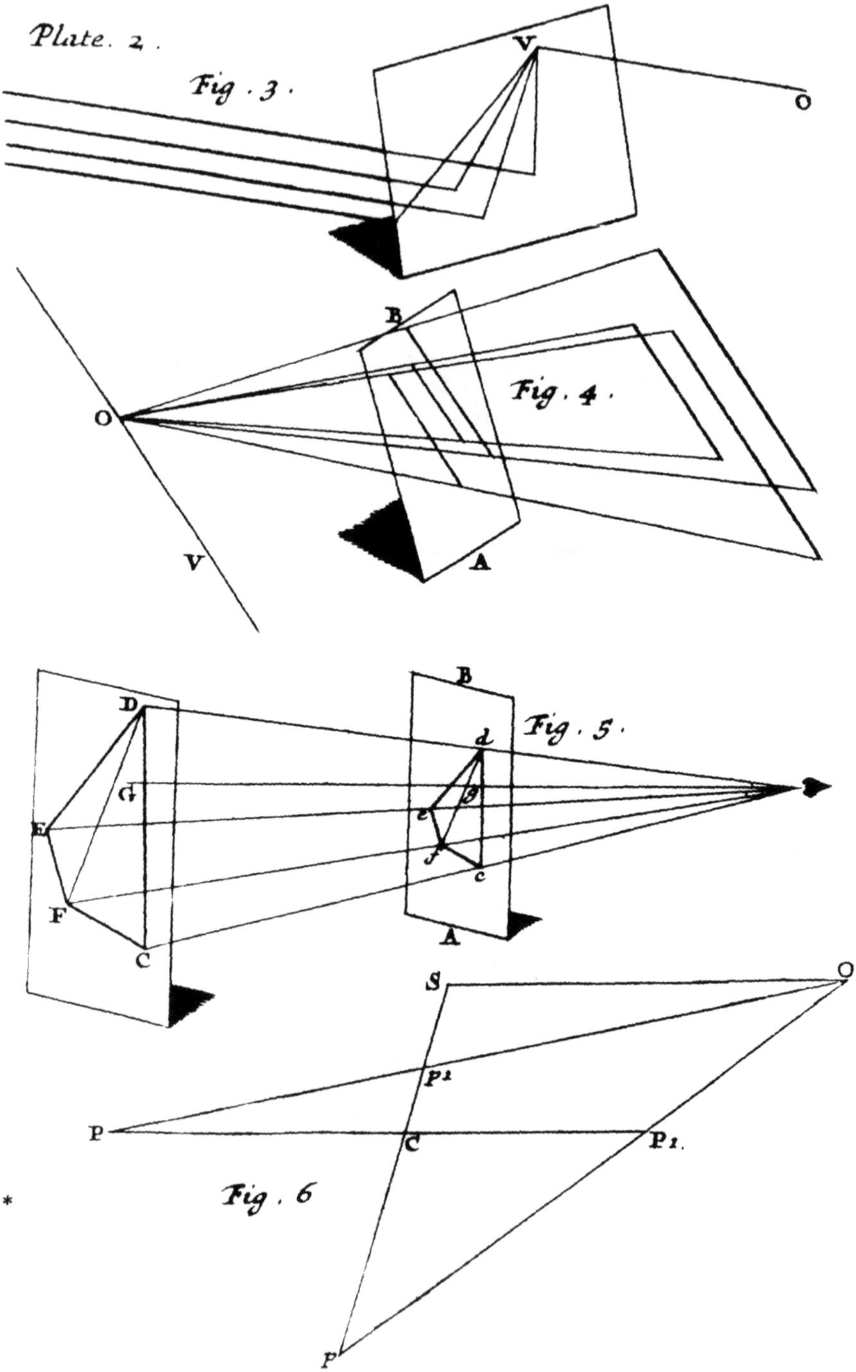
Plate. 2.
Fig. 3.
V
O
B
Fig. 4.
O
V
A
B
D
Fig. 5.
d
G
g
E
e
f
F
c
A
C
S
O
P1
P
C
P1.
Fig. 6
P

*

Plate . 3 .

Fig . 7 .

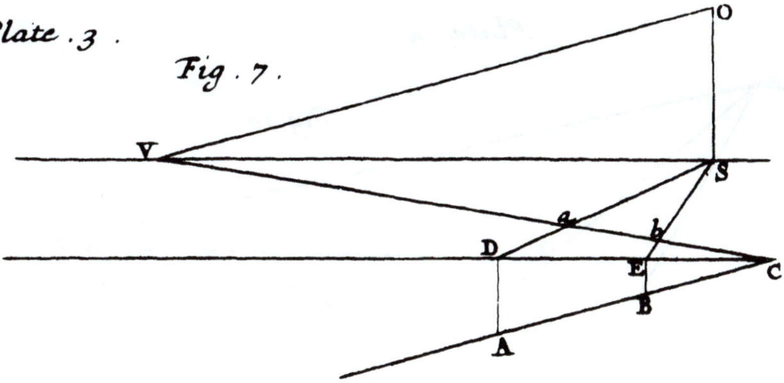

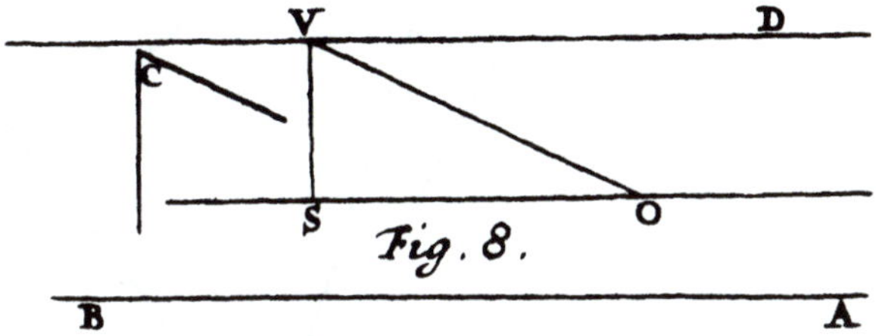

Fig. 8.

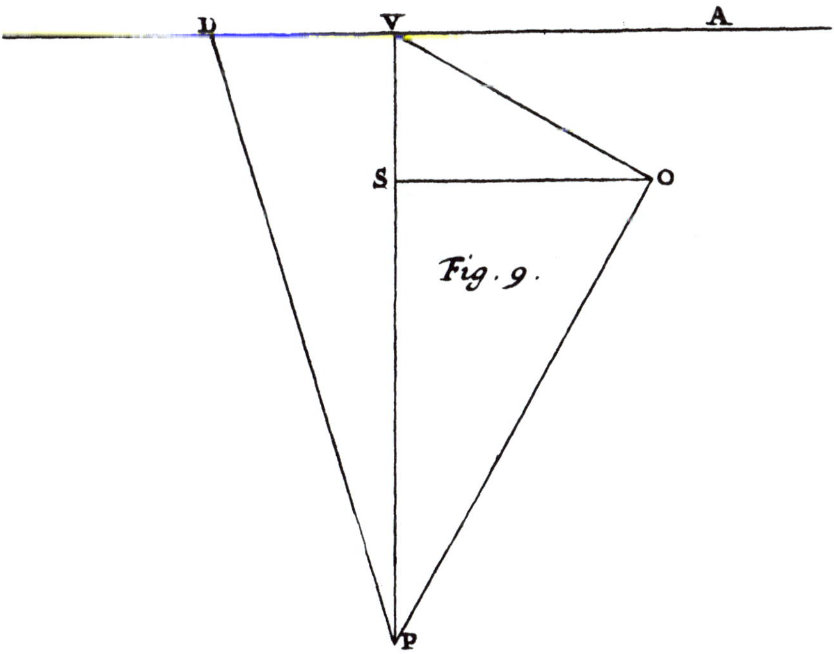

Fig. 9.

Plate . 4 .

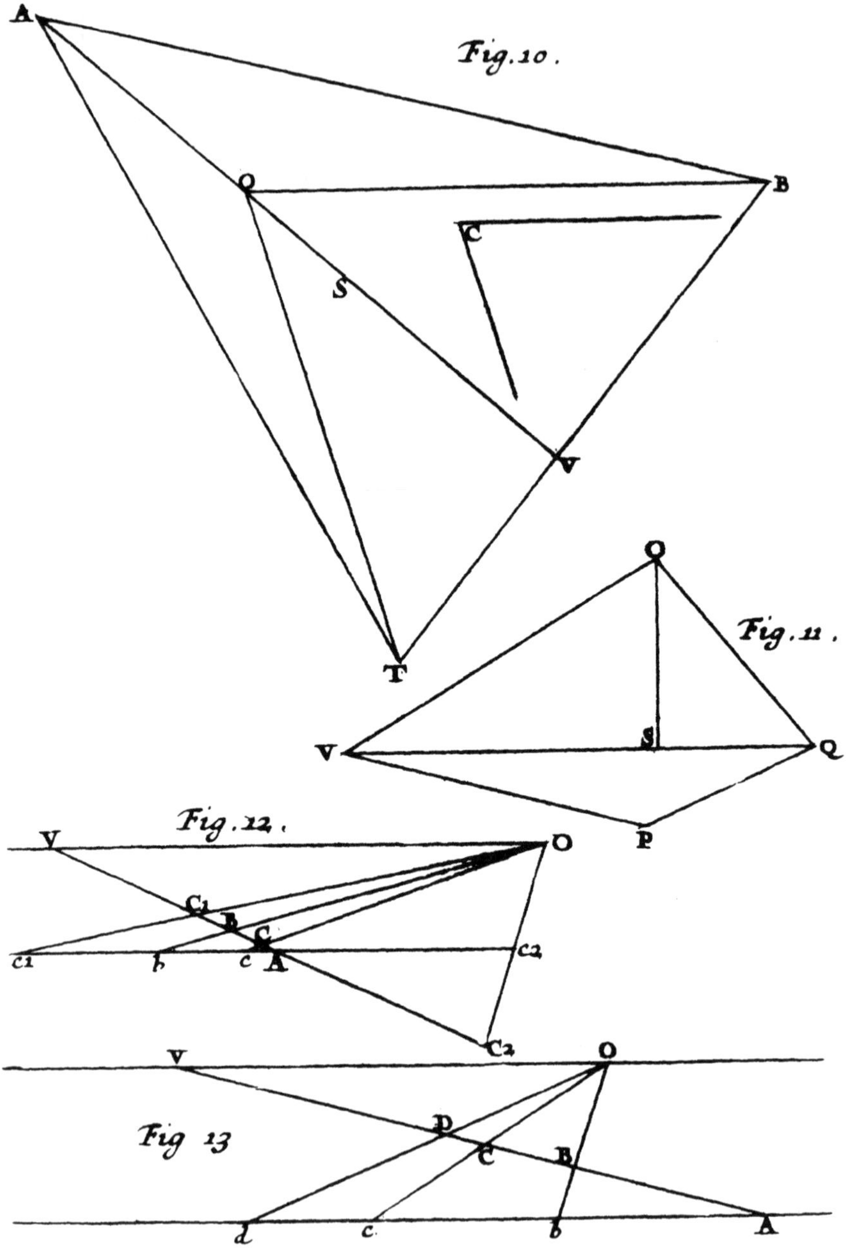
Fig.10 .
A
O
B
C
S
V
T
Fig.11 .
O
V
S
Q
P
Fig.12 .
V
O
C1
B
C
c1
b
c
A
c2
C2
V
O
Fig 13
D
C
B
d
c.
b
A

Plate. 5.

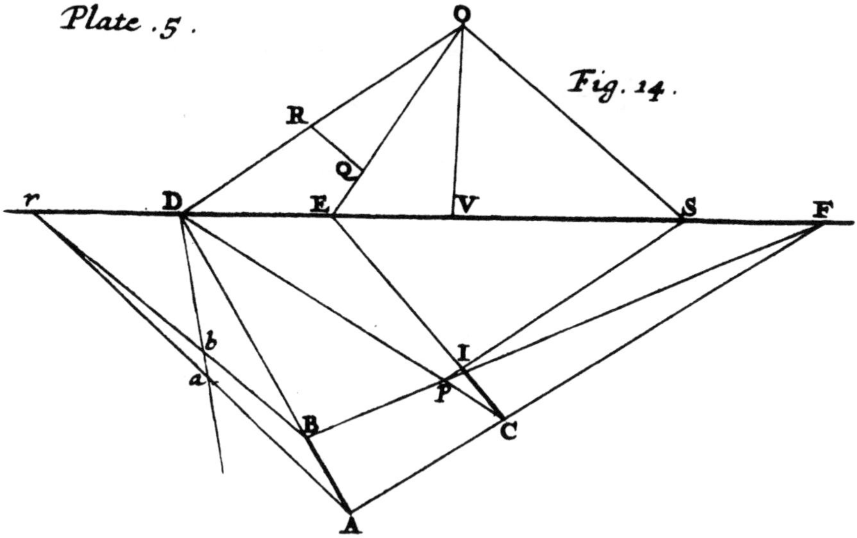

Fig. 14.

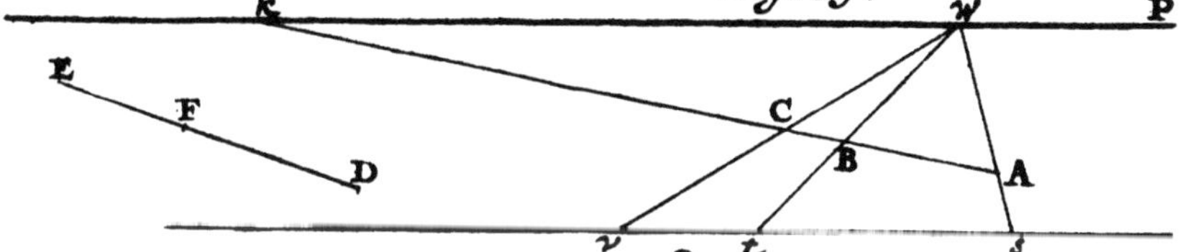

Fig. 15.

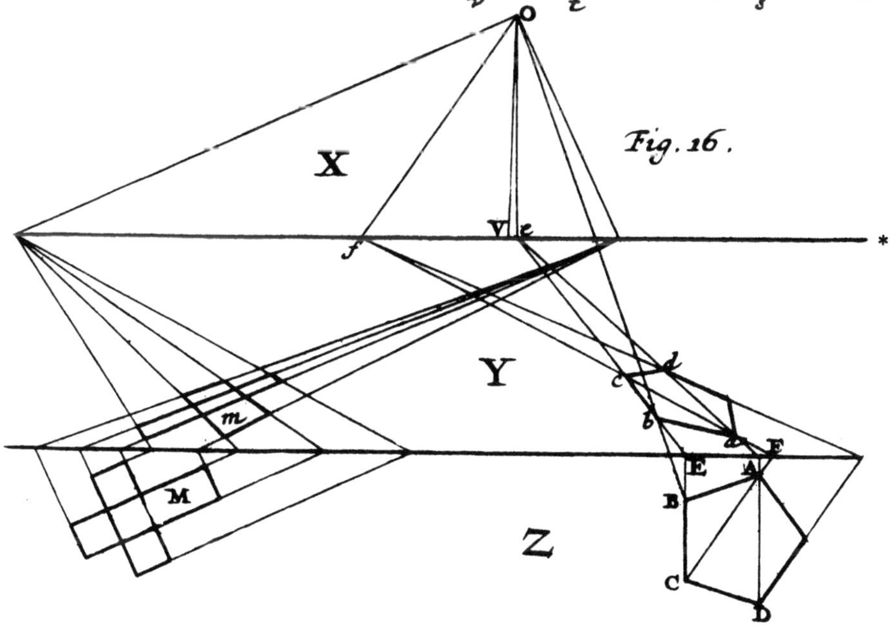

Fig. 16.

Plate .6 .

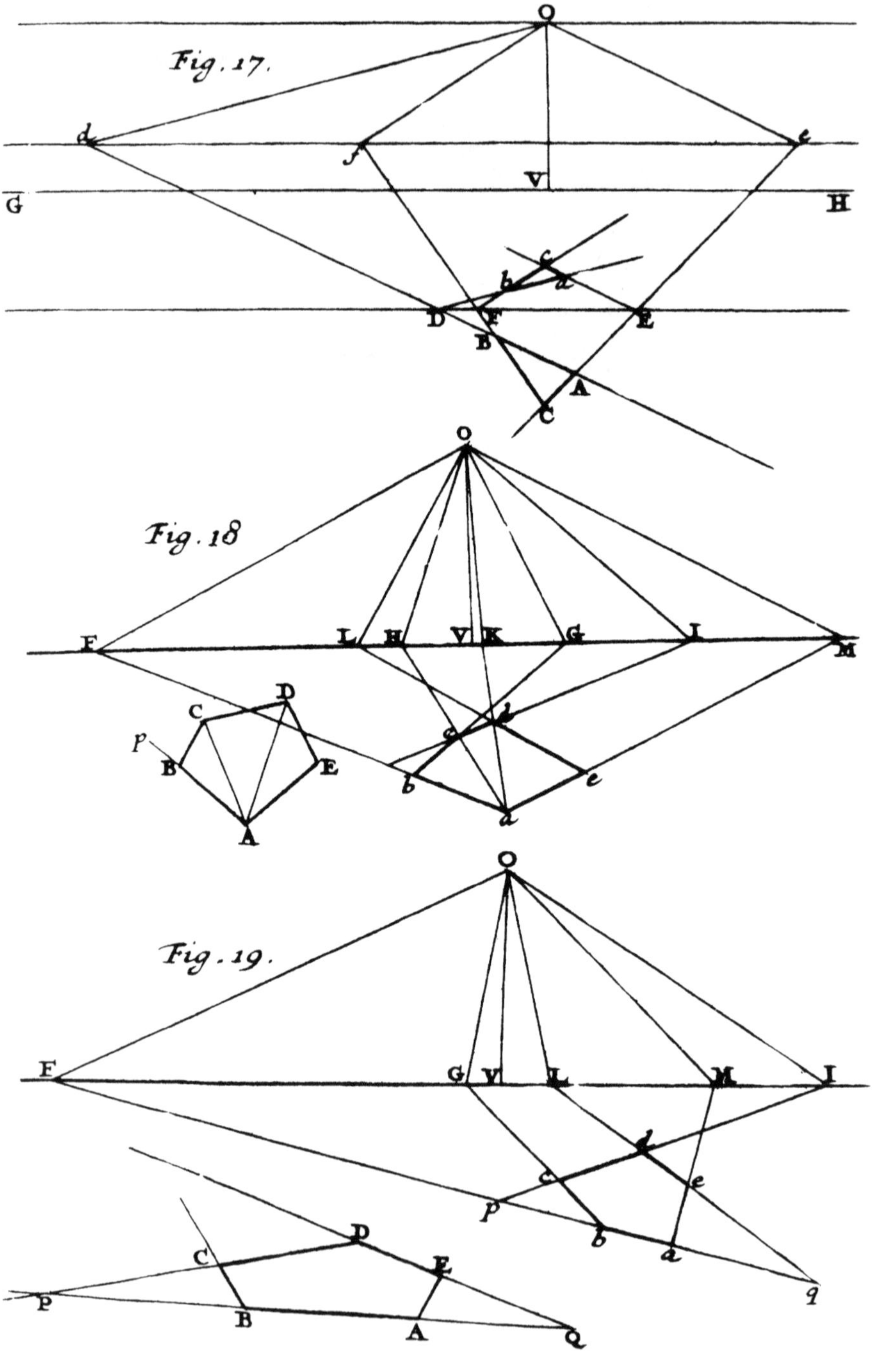
Fig. 17.
O
d
f
e
V
G
H
c
b
a
D
F
E
B
A
C
Fig. 18
O
F
L
H
V
K
G
I
M
D
C
p
B
E
A
d
c
b
e
a
Fig. 19.
O
F
G
V
L
M
I
d
c
e
p
b
a
q
D
C
E
P
B
A
Q

Plate. 7.

Fig. 20.

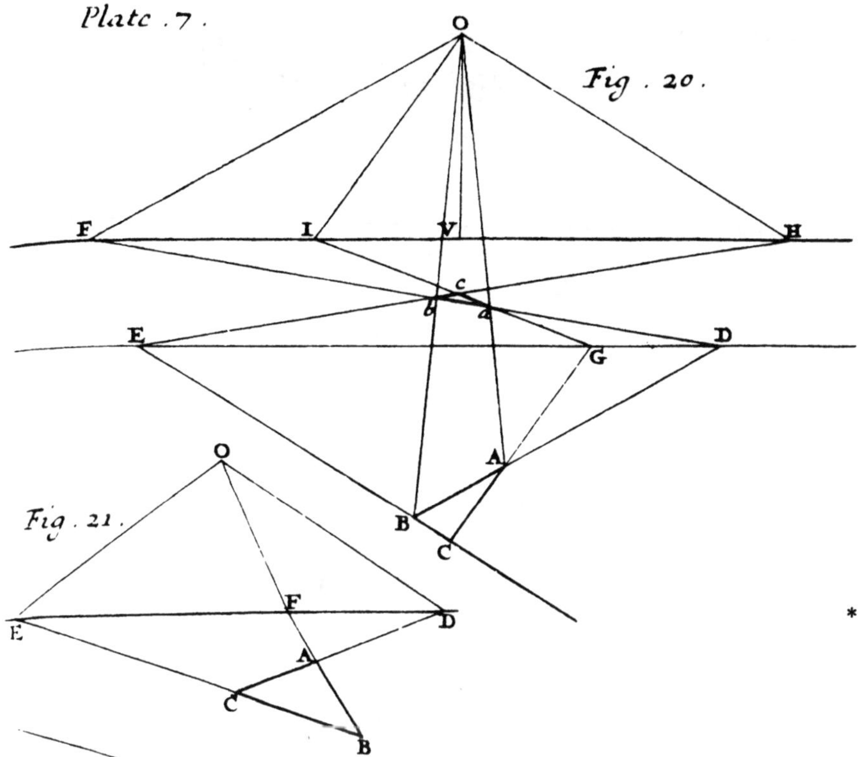

*

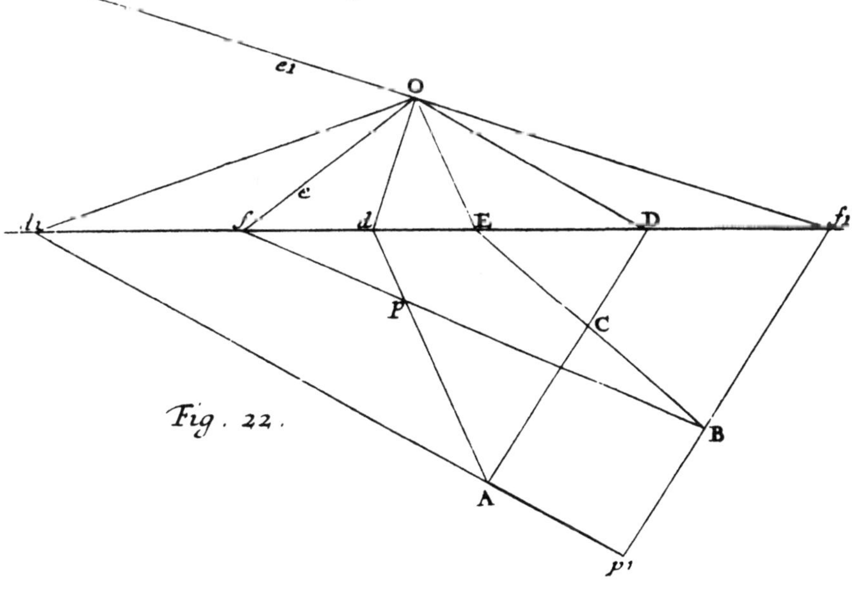

Fig. 22.

Plate. 8

Fig 23.

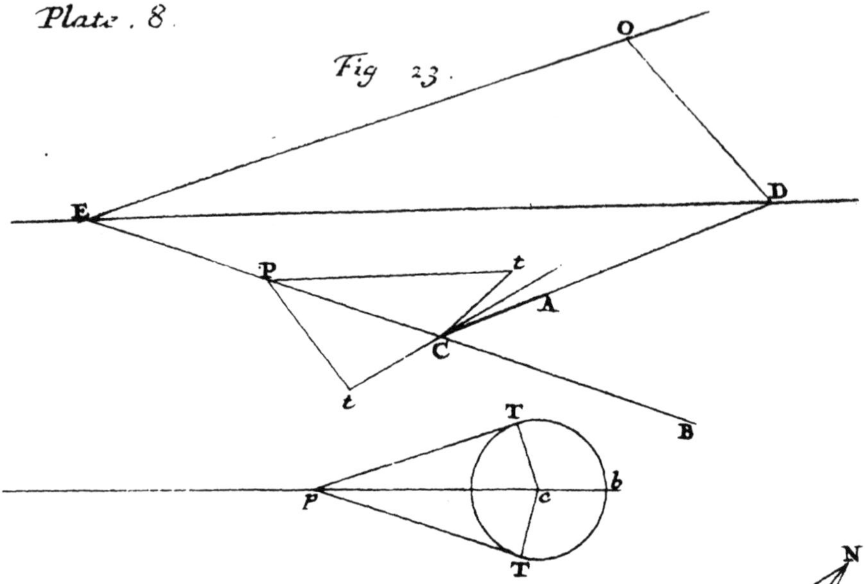

Fig 24.

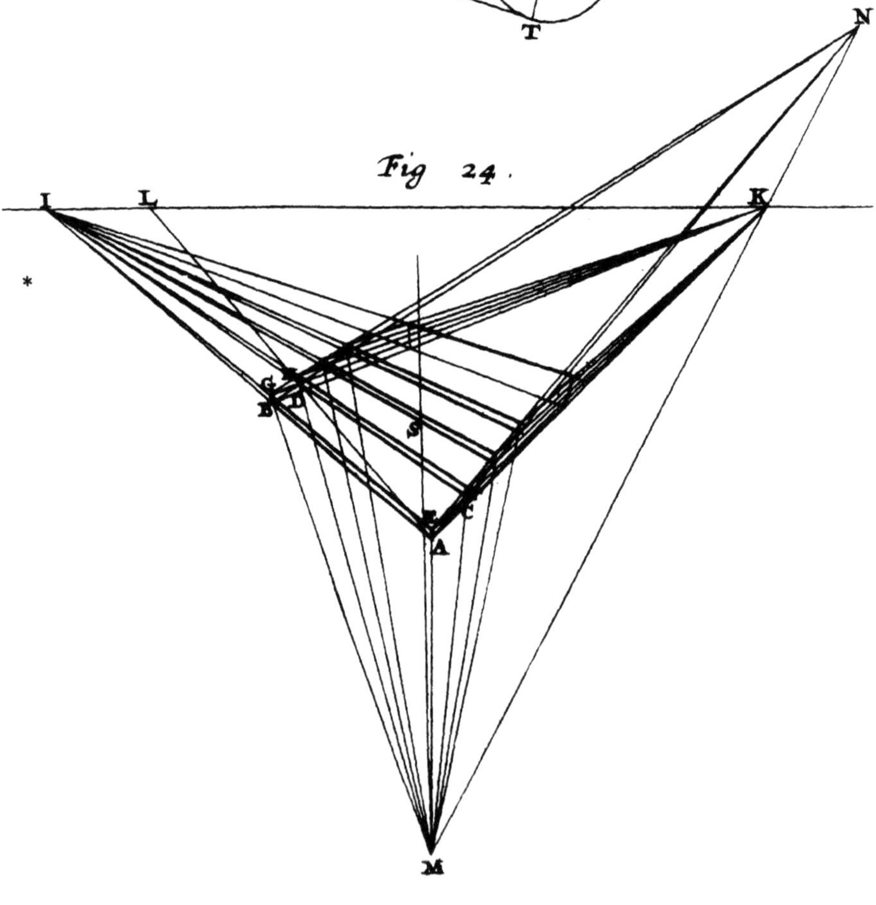

Plate. 9.

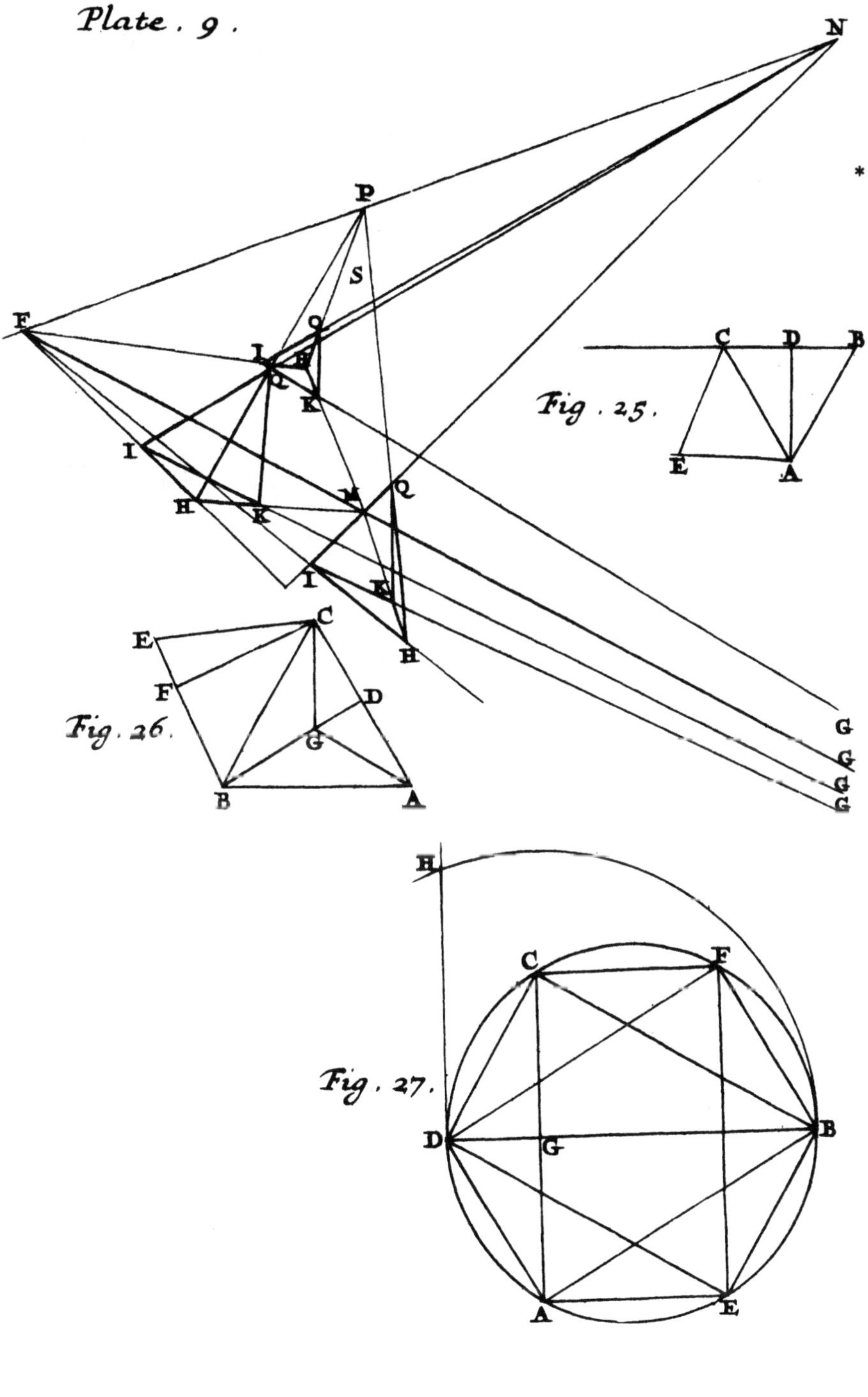

Plate . 10 .

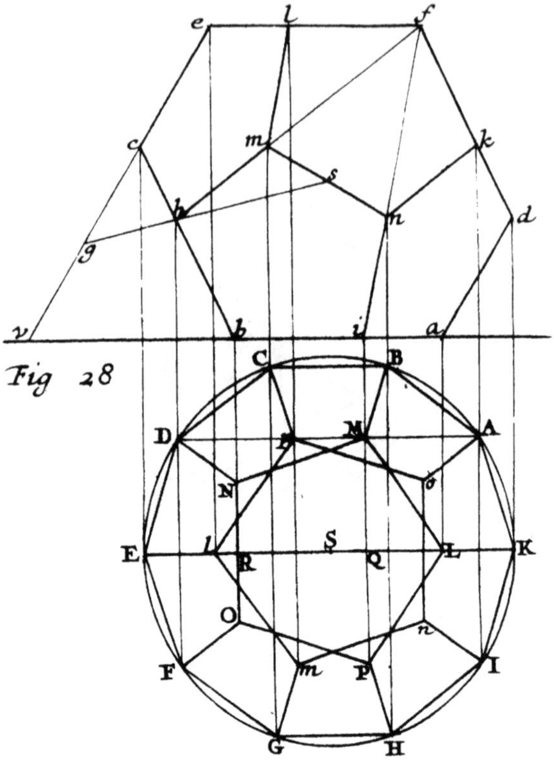

Fig 28

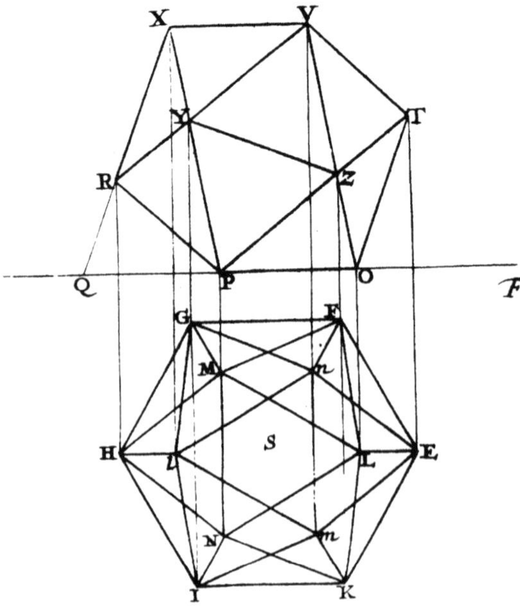

Fig . 29 .

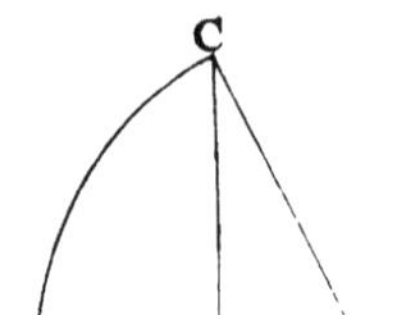

Plate. 11.

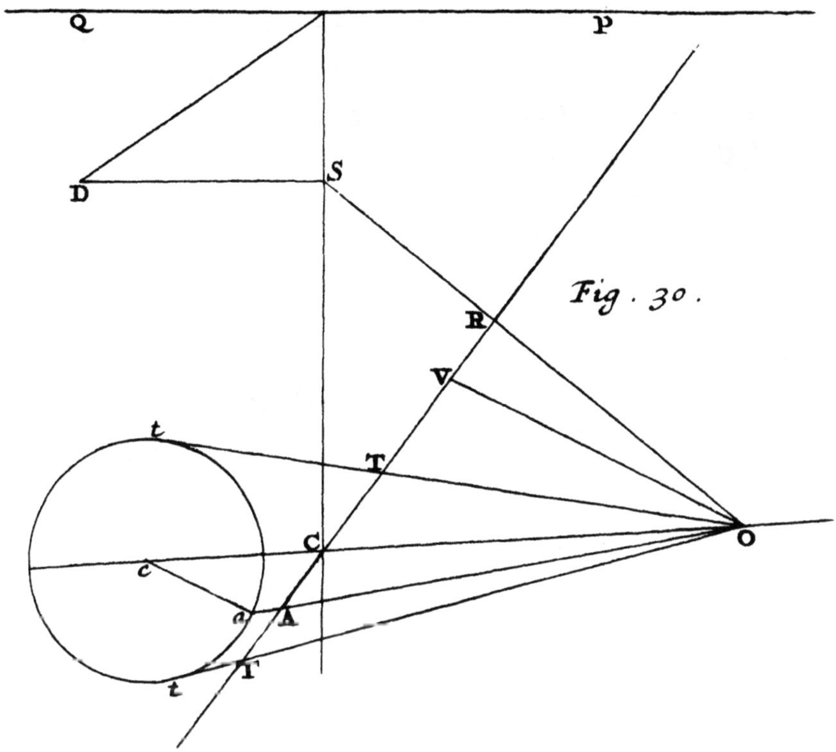

Fig. 30.

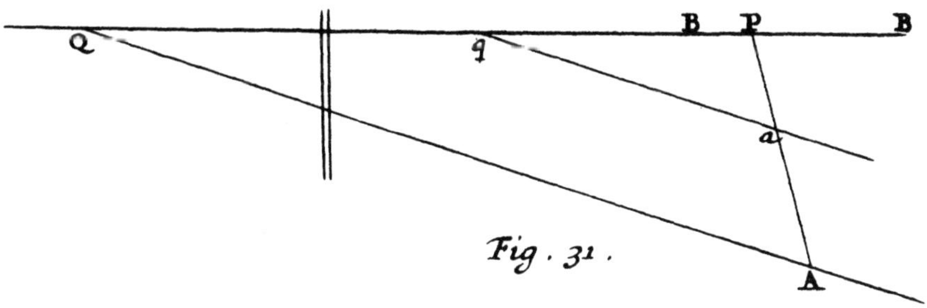

Fig. 31.

Plate . 12.

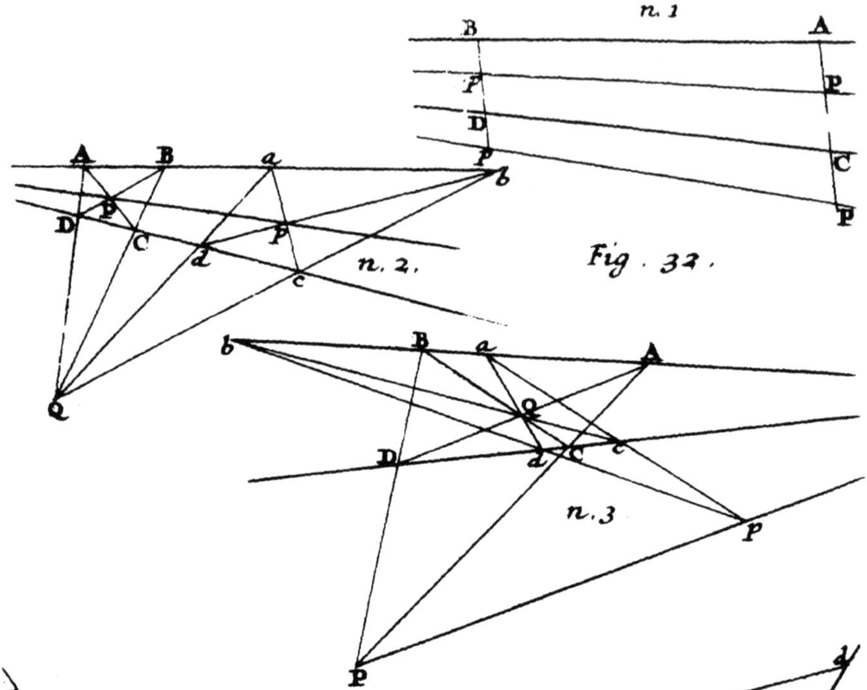

Fig . 32.

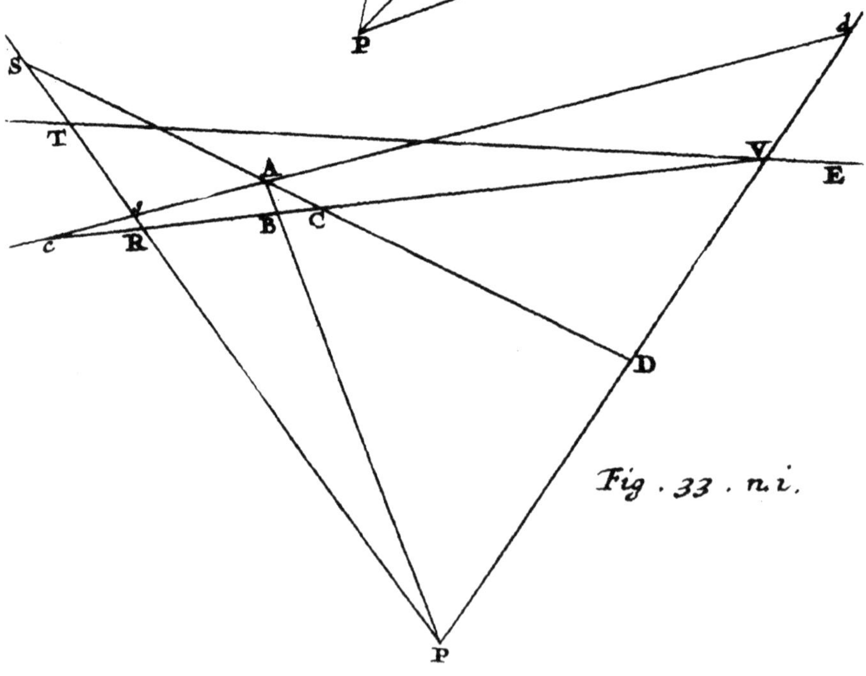

Fig . 33 . n.1.

Plate. 13.

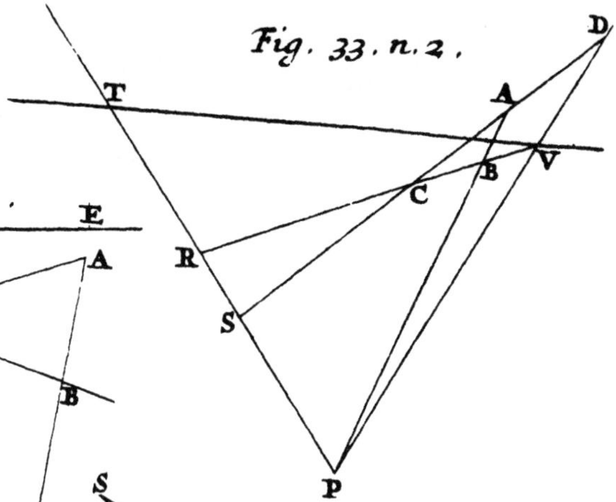

Fig. 33. n. 2.

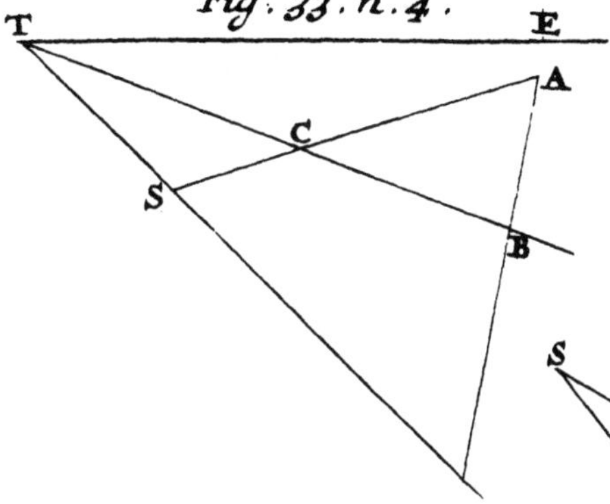

Fig. 33. n. 4.

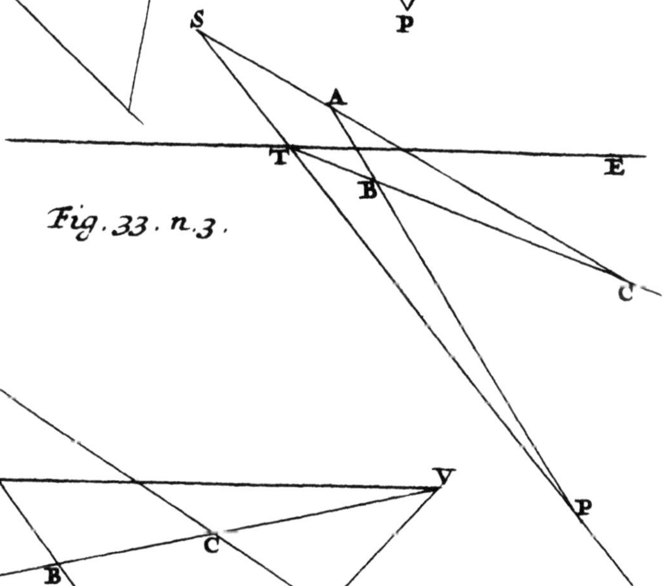

Fig. 33. n. 3.

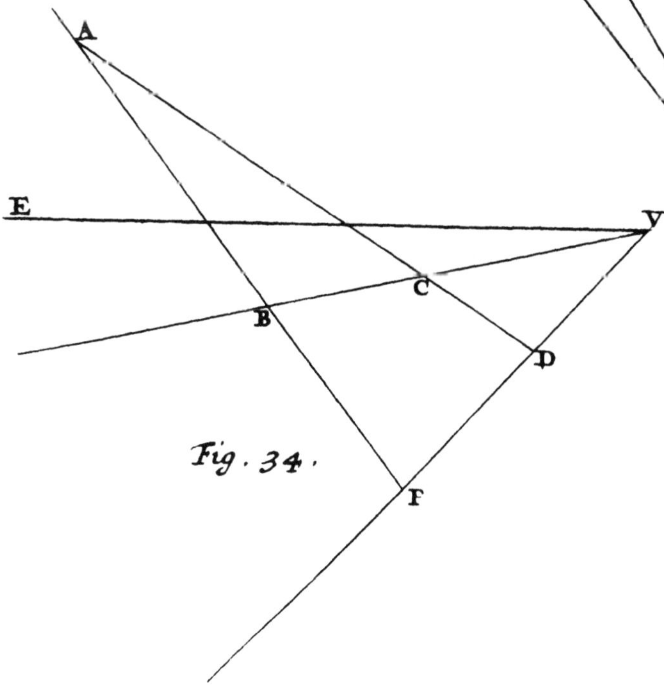

Fig. 34.

Plate. 14.

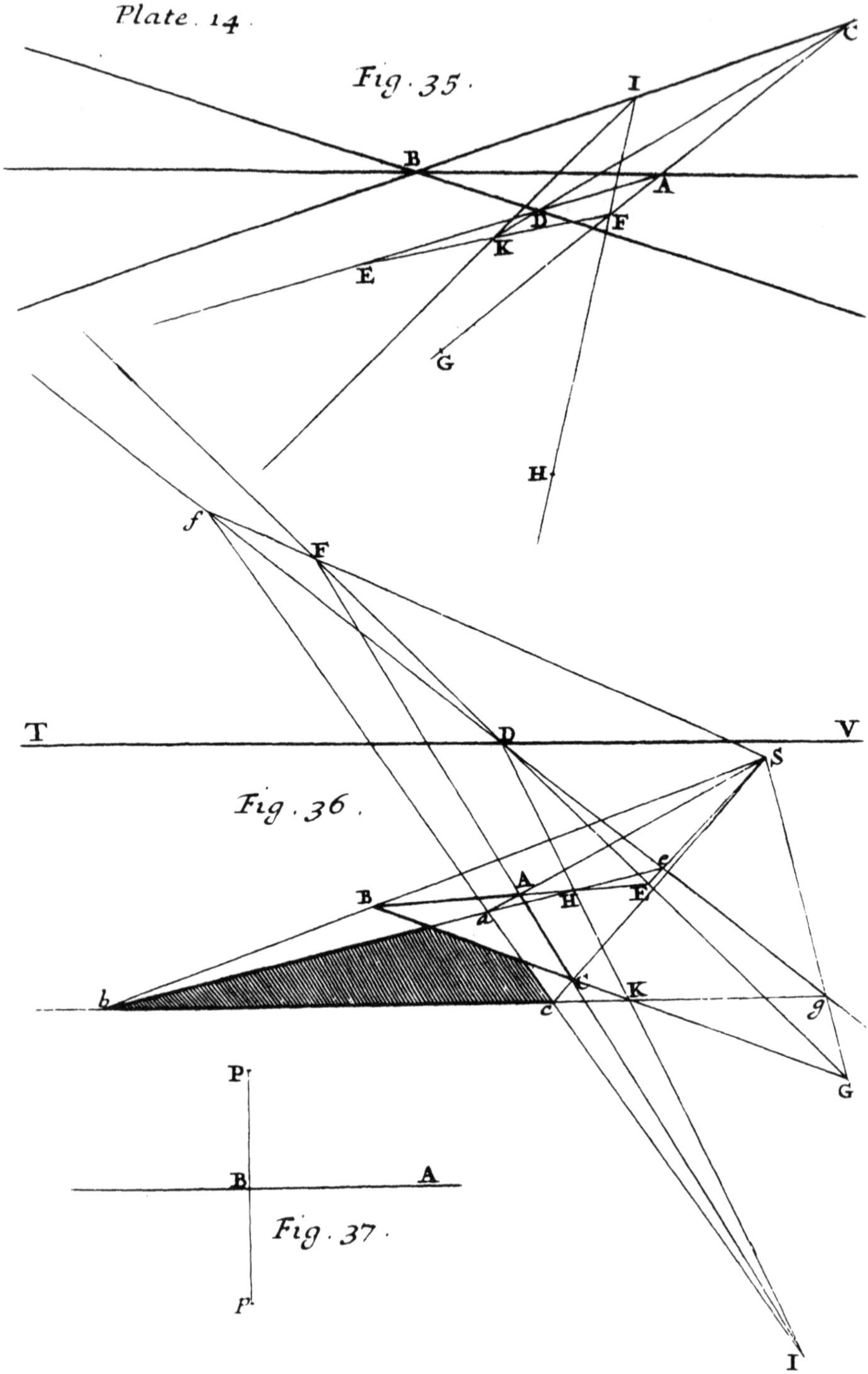

Plate. 15.

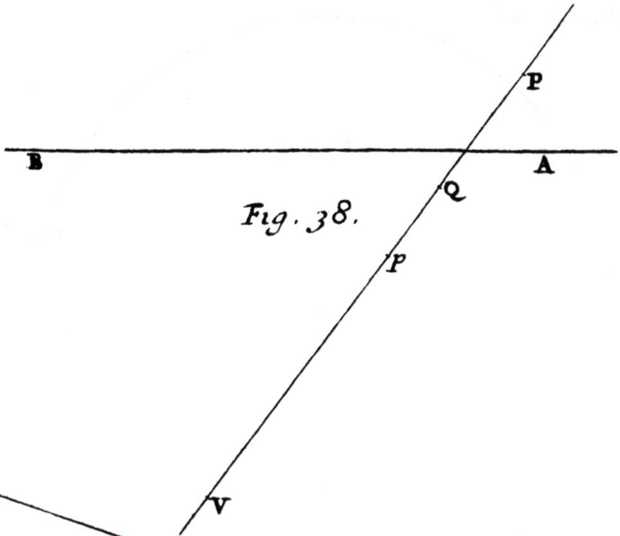

Fig. 38.

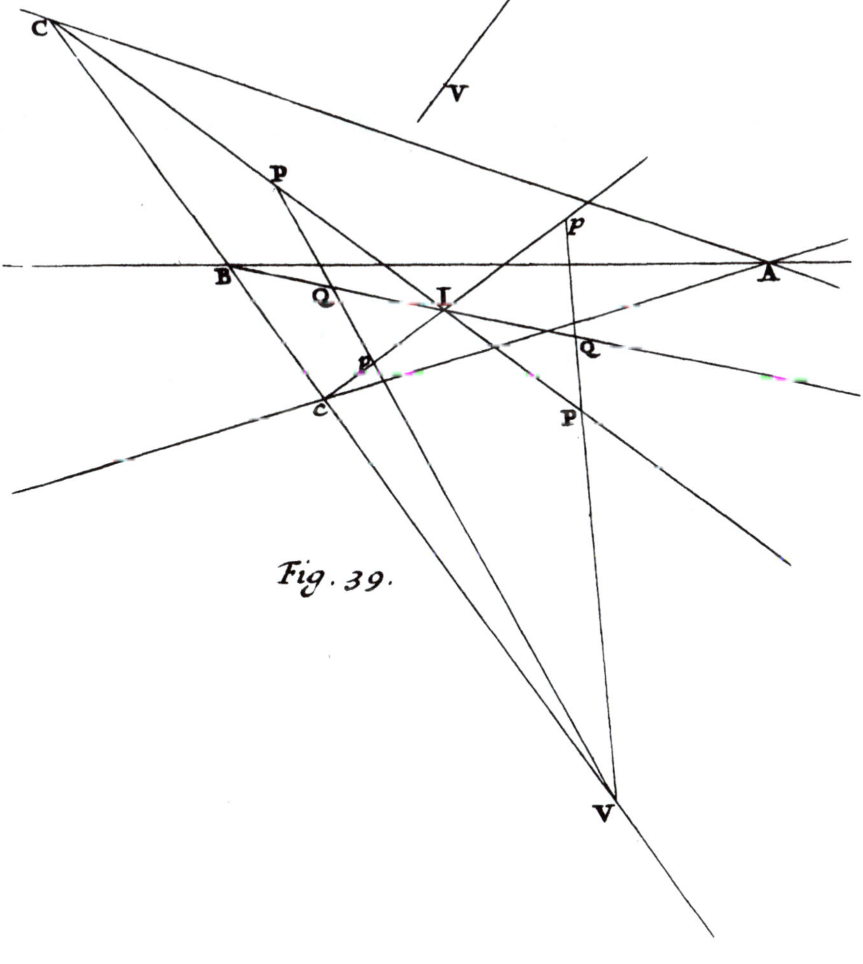

Fig. 39.

Plate. 16

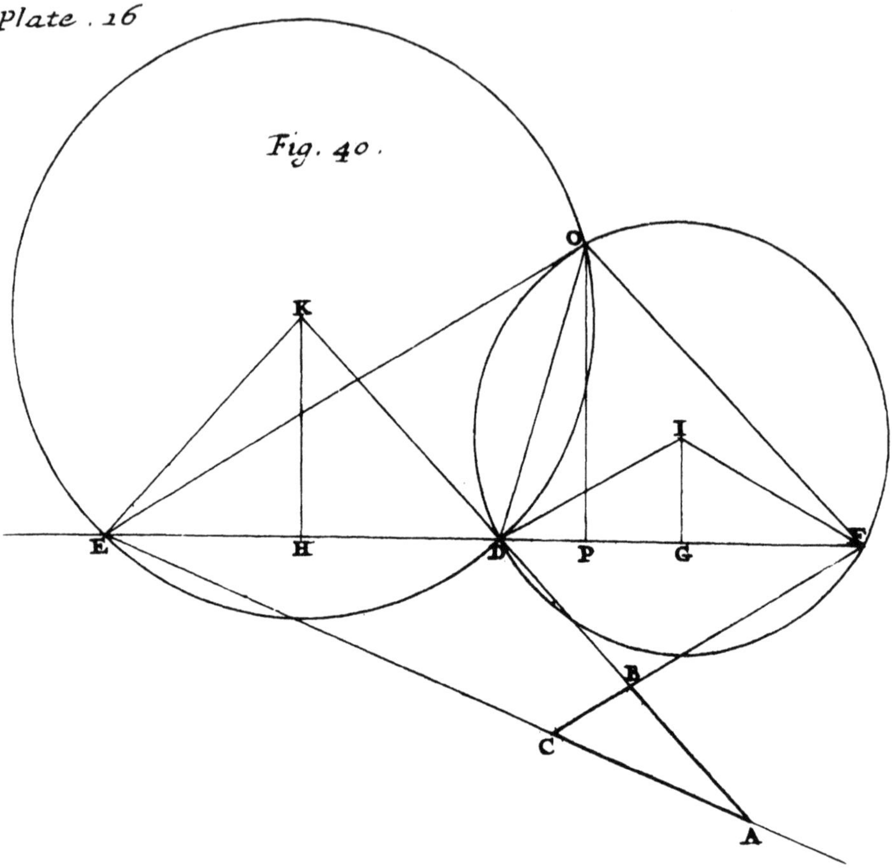

Fig. 40.

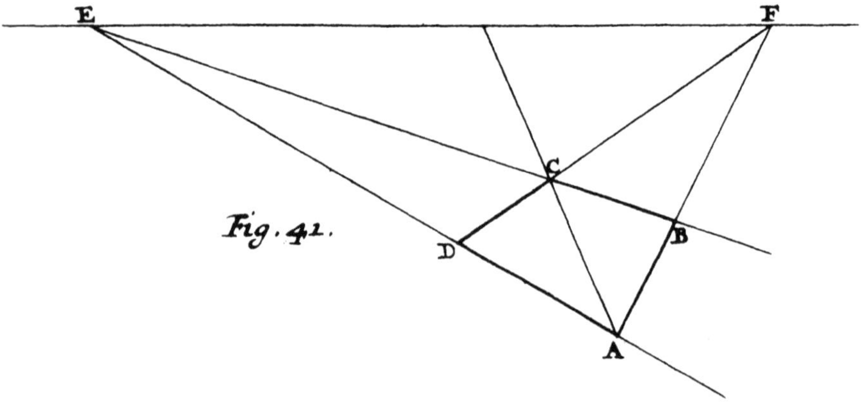

Fig. 41.

Plate. 17.

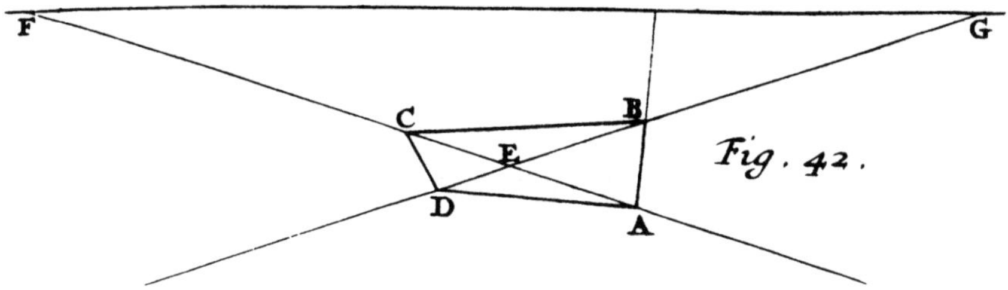

Fig. 42.

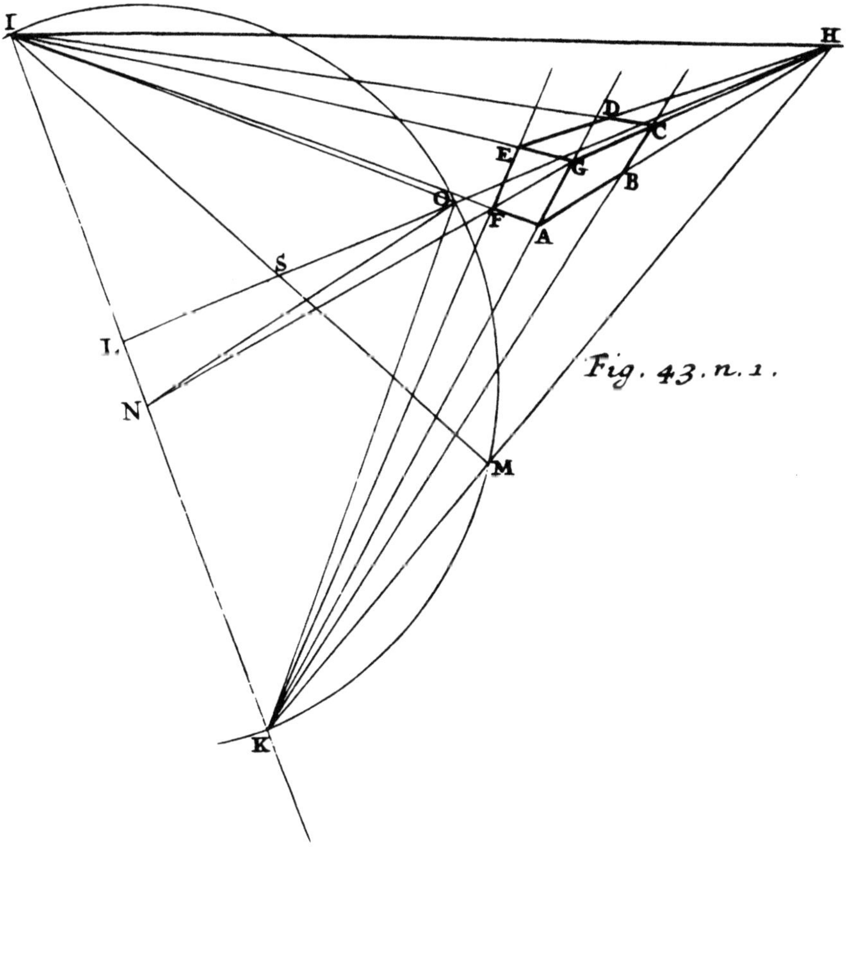

Fig. 43. n. 1.

Plate 18.

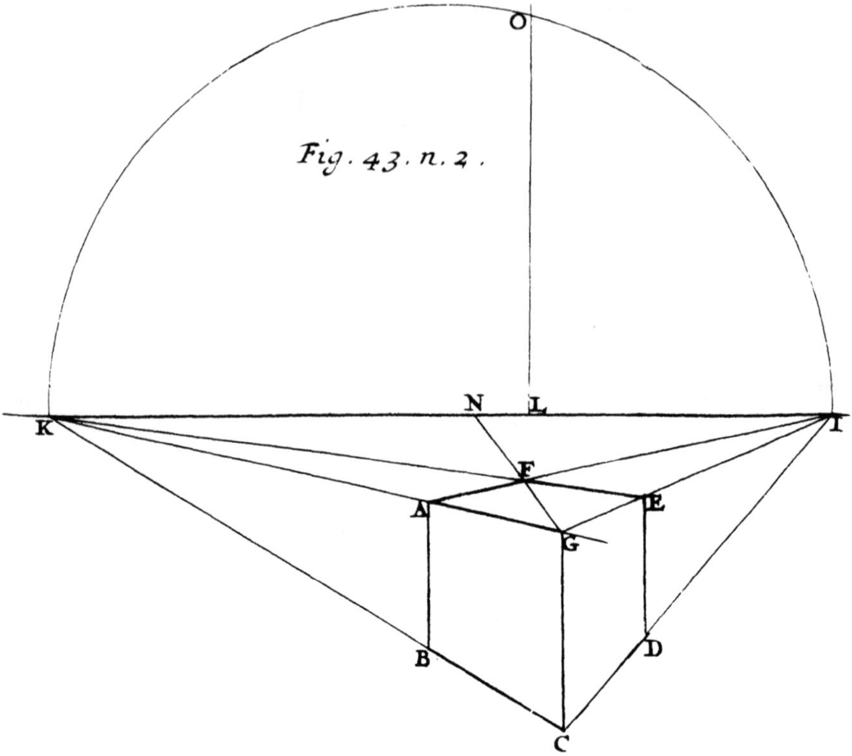

BOOKS Printed for R. Knaplock, *at the* Bishop's Head *in St.* Paul's *Church-Yard.*

AN Attempt towards Recovering an Account of the Numbers and Sufferings of the Clergy of the Church of *England*, Heads of Colleges, Fellows, Scholars, *&c.* who were Sequester'd, Harrass'd, *&c.* in the late Times of the Grand Rebellion: Occasioned by the Ninth Chapter (now the Second Volume) of Dr. *Calamy*'s Abridgment of the Life of Mr. *Baxter*. Together with an Examination of that Chapter. By *John Walker*, D. D. Rector of St. *Mary the More* in *Exeter*, and sometime Fellow of *Exeter* College in *Oxford*. *Folio*. Price 23 *s*.

The *English* Historical Library in Three Parts. Giving a short View and Character of most of our Historians, either in Print or Manuscript. With an Account of our Records, Law-Books, Coins and other Materials. Serviceable to the Undertakers of a General History of *England*. The Second Edition Corrected and Augmented by *W. Nicholson*, Archdeacon (now Bishop) of *Carlile*. In *Folio*. Price 10 *s*.

M. Fabij Quintiliani de Institutione Oratoria Libri Duodecim, Juxta Editionem, quæ ad Fidem trium Codicum M S S. & Octo Impressorum, prodiit è Theatro Sheldoniano Oxonii 1693. *Huic Editione accedunt Notæ Maximæ & Necessariæ è Turnebo & aliis, per Edm. Gibson, D. D.* In Octavo. Price 7 *s*.

Origines Ecclesiasticæ: Or, The Antiquities of the Christian Church. Vol. I. In Two Books. Whereof the first Treats of Christians in general; their several Names and Degrees; of *Catechumens*, Laity and Clergy. And the second gives an Account of the several Superior Orders and Offices in the Primitive Church. *The Second Edition.*

Origines Ecclesiasticæ: Vol. II. Giving an Account, 1. Of the Inferior Orders of the Ancient Clergy. 2. Of the manner of their Elections and Ordinations, and the particular Qualifications of such as were to be Ordain'd. 3. Of their Privileges, Immunities and Revenues. 4. Of the several Laws relating to their Employment, Life and Conversation. *The Second Edition.*

Origines

Books Printed for R. Knaplock..

Origines Ecclesiasticæ. Vol. III. Wherein is contain'd, 1. An Account of the Ancient Asceticks, and the Original of Monks succeeding them, with the several Laws and Rules relating to the Monastick Life. 2. An Account of the Ancient Churches, their Originals, Names, Parts, Utensils, Consecrations, Immunities, &c. 3. A Geographical Description of the Districts of the Ancient Church, on Account of its Division into Provinces, Dioceses, and Parishes, and of the first Original of these. The whole Illustrated with Cuts, Maps, and Indexes.

Origines Ecclesiasticæ: Or, The Antiquities of the Christian Church. Vol. IV. In Three Books. Giving an Account 1. Of the Institution of the *Catechumens*, and the first Use of Creeds in the Church. 2. Of the Rites and Customs observ'd in the Administration of Baptism. 3. Of Confirmation, and other Rites following Baptism, before Men were made Partakers of the Eucharist. By *Joseph Bingham*, Rector of *Havant.*

A New View of *London:* Or, An Ample Account of that City. In Two Volumes. Being a more particular Description thereof than has hitherto been publish'd of any City in the World. 1. Containing the Names of the Streets, Squares, Lanes, Markets, Courts, Alleys, Rows, Rents, Yards, and Inns in *London*, *Westminster* and *Southwark*, &c. 2. Of the Churches; their Names, Building, Ornament, Dimensions, Benefactors; Monuments, Tombs, Cenotaphs, &c. describ'd; with their Epitaphs, Inscriptions, Motto's, Arms, &c. 3. Of the several Companies; their Nature, Halls, Armorial Ensigns Blazoned, &c. 4. Of the King's Palaces, Eminent Houses, &c. of the Nobility, House of Lords and Commons, &c. 5. Colleges, Libraries, Musæums, Repositories, Free-Schools, &c. 6. The Hospitals, Prisons, Work-Houses, &c. 7. Of Fountains, Bridges, Conduits, Ferries, Docks, Keys, Wharfs, Plying Places for Boats, &c. 8. An Account of about 90 publick Statues. Their Situations, Descriptions, &c. To the whole is prefix'd, *An Introduction* concerning *London* in general; its Antiquity, Magnitude, Walls and Gates, Number of Houses, Inhabitants, Males, Females, Fighting Men; its Riches, Strength, Franchises, Government, Civil, Ecclesiastical, and Military, &c. Illustrated with two Plans, *viz.* 1. Of *London*, as in Q. *Elizabeth*'s Time. 2. As it is at present.

Notes to *Linear Perspective*

(by Kirsti Andersen)

The references are to page numbers of this book, the original page numbers are indicated in parentheses. A minus after a line number signifies that the counting is made from the bottom of the page.

1. p. 76 (2), l. 11: For *ET* read *EF* (cf. Errata, p. 72).
2. p. 77 (3), l. 15: For *AP* read *AB* (cf. Errata, p. 72).
3. p. 80 (6), l. 4: For in read is.
4. p. 81 (7): Theorem 1, cf. p. 12 and Theorem III, p. 174.
5. p. 81 (7), l. 1^-: In *New Principles*, p. 176, Taylor proved this theorem in a more traditional way.
6. p. 82 (8), l. 15: For *cdf* read *edf* (cf. Errata, p. 72).
7. p. 82 (8), l. 16: For *edcf* read *cdef* (cf. Errata, p. 72).
8. p. 83 (9): Theorem 3, cf. Theorem V, p. 177.
9. p. 84 (10): Problem 1, cf. pp. 13–14 and Problem I, p. 180.
10. p. 84 (10), l. 9: By seat Taylor meant orthogonal projection, cf. p. 165.
11. p. 85 (11): Problem 2, cf. Problem II, p. 181.
12. p. 85 (11), l. 5^-: It is assumed that the distances between the two original points and the picture are given.
13. p. 86 (12): Problem 3, cf. Problem V, p. 185.
14. p. 87 (13): Problem 4, cf. pp. 34–36 and Problem XIV, p. 197.
15. p. 87 (13), l. 1^-: For *OP* read *SP*.
16. p. 88 (14): Problem 5, cf. pp. 34, 37 and Problem XV, p. 199.
17. p. 89 (15): Problem 6, cf. Problem XVI, p. 200.
18. p. 90 (16): Problem 7, cf. pp. 37–38 and Problem XVII, p. 201.
19. p. 91 (17): Problem 8, cf. pp. 26–27, 30 and Problem XI, p. 192.
20. p. 92 (18): Problem 9, cf. pp. 26–30 and Problem III, p. 183.
21. p. 92 (18), ll. 11–12: The relation

$$AC : AB = (Ac \cdot AV) : (Ac \cdot AV + bc \cdot BV)$$

is equivalent to

$$AC : CB = (Ac \cdot AV) : (bc \cdot BV),$$

Taylor used the latter in *New Principles* (p. 183), how it can be derived is shown in note 10, p. 28.

22. p. 93 (19): Problem 10, cf. pp. 26–27 and Problem IV, p. 184.
23. p. 94 (20), l. 11^-: For vanishing point *FD* read vanishing line *FD*.
24. p. 96 (22): Example I, cf. Problem VI and Problem VII, pp. 186–189.
25. p. 96 (22), ll. 11–21: Here Taylor applied the construction I term "the visual ray construction," for a description, see pp. 16–19.
26. p. 97 (23), l. 1: For *DT* read *DF* (cf. Errata, p. 72).
27. p. 97 (23): Example II, cf. Problem XIII, p. 194.
28. p. 98 (24), l. 1^-: For *OD* read *Od*.
29. p. 100 (26): Example VII, cf. Problem XVIII, p. 203.
30. p. 101 (27), l. 15 to p. 102 (28), l. 3: To make a plan of the dodecahedron Taylor projected it upon one of its faces: the one which is denoted *LMNOP* in Figure 4 (p. 141). As the plane of elevation—which will here be called η—he chose the plane which is perpendicular to *LMNOP* and contains the line *LR*, the latter being normal to *NO*. The plane then also contains the center, *s*, of the dodecahedron.

    Taylor's construction of the plan and elevation builds upon many results concerning a dodecahedron and its pentagonal faces. In the following I shall list and deduce some of these results. Let us first consider a regular pentagon *ABCEF* (Figure 1); let *AD* be normal to *CE*, *O* the center of the circumscribed circle, and

$$AB = \sigma, \qquad AC = \delta, \qquad AD = \lambda, \qquad OC = \rho.$$

A determination of the relevant angles shows that triangle *BHC* is isosceles, which implies that $HC = \sigma$ and $AH = \delta - \sigma$; it furthermore shows

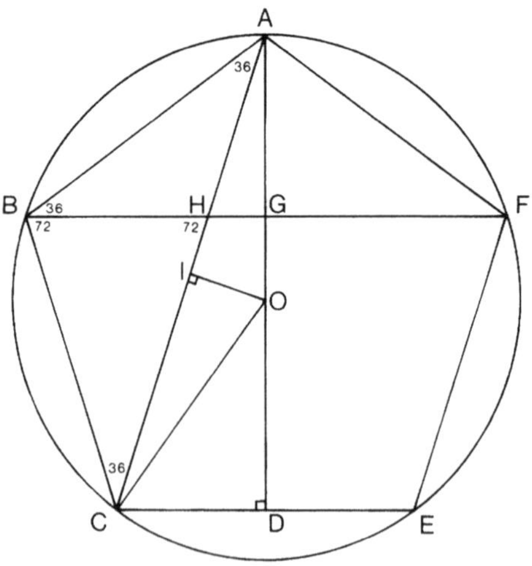

Figure 1. A pentagon.

that the triangles ABC and ABH are similar and hence

$$\delta : \sigma = \sigma : (\delta - \sigma), \tag{1}$$

which is equivalent to

$$\delta : (\sigma + \delta) = \sigma : \delta. \tag{2}$$

When I is the midpoint of AC the triangles ACD and OIC are also similar and hence

$$\delta : (2\lambda) = \rho : \delta. \tag{3}$$

We proceed by considering the polygon in which η cuts the dodecahedron (it is advisable to look at a three-dimensional dodecahedron). Some geometrical observations show that it is the hexagon depicted in Figure 2; in this the sides ce and ad are original sides of the dodecahedron, the point f is the midpoint of one of the sides of the top pentagon (that which in Figure 4 is depicted as **no**) and similarly b is the midpoint of one of the sides of the bottom pentagon (b is the same point as R in Figure 4). Thus we have (compare with Figure 1) that

$$ab = bc = ef = fd = \lambda, \qquad \text{and} \qquad ad = ce = \sigma. \tag{4}$$

It can, furthermore, be noticed that the side ec is parallel to the diagonal in an opposite pentagon and that

$$cv = \delta. \tag{5}$$

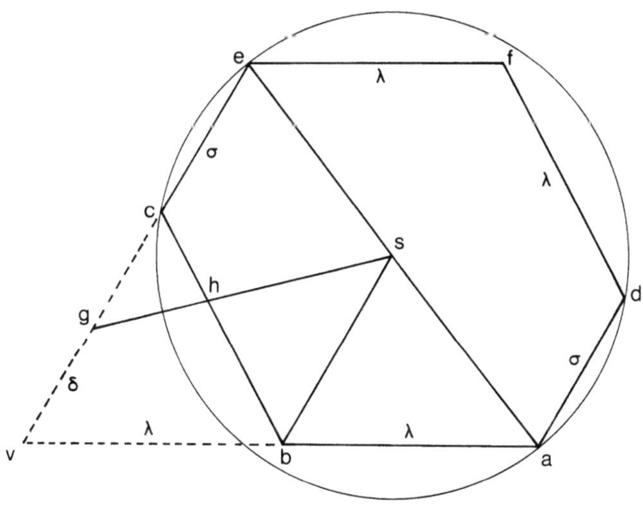

Figure 2. A section of a dodecahedron.

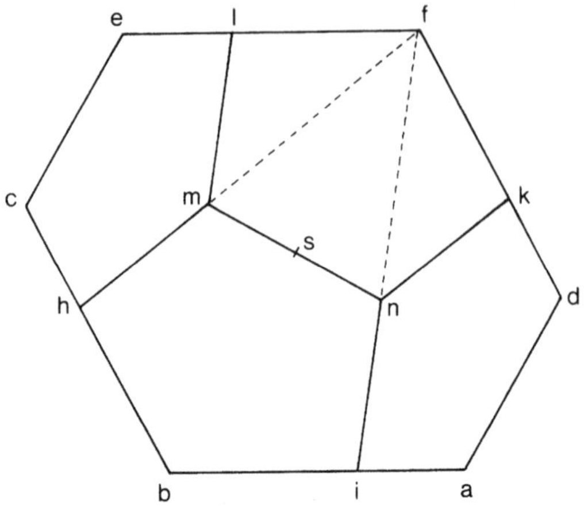

Figure 3. A diagram similar to the upper part of Taylor's Figure 28 (p. 126).

The length of bv can also be determined: If we let $< abc = 2\Theta$, then a calculation of the angles in the hexagon shows that $< bce = 180° - \Theta$ which means that the triangle bcv is isosceles and

$$bv = \lambda. \tag{6}$$

By applying the results (4)–(6) and the facts that line *ef* is parallel to *ab* and the line *fd* to *cb* Taylor constructed the points *a*, *b*, *c*, *e*, *f*, *d* of the elevation. To determine the projections, *h*, *i*, *k*, *l*, *m*, *n*, of the remaining vertices of the dodecahedron Taylor, explicitly and implicitly, used a collinearity of some points: in his Figure 28 (p. 126), partly reproduced in Figure 3 above, the points *b*, *h*, *c*, for instance, lie on a line, as do *h*, *m*, *f*; and also *i*, *n*, *f*. That this is correct can be realized by observing that the original of line **no** (Figure 4), and all diagonals in the pentagonal faces parallel to it, are perpendicular to η. This orthogonality also implies that the points *h*, *i*, *k*, *l* (Figure 3) divide the line segments *cb*, *ba*, *df*, *fe* in the same ratio as the point *H* in Figure 1 divides *AC*. In other words (cf. (1)).

$$hc : hb = ia : ib = kd : kf = le : lf = (\delta - \sigma) : \sigma = \sigma : \delta. \tag{7}$$

Since (cf. (4)) $cb = ba = df = fe$, the relation (7) in particular implies that

$$hb = ib = kf = lf. \tag{8}$$

Taylor applied the relation (8), but he did not use (7) to find the length of *hb*. Instead he based his construction on the following observation: Let *g* be the point of intersection of *sh* and *cv* (Figure 2), then—as will be shown—

$$gv = gc. \tag{9}$$

To obtain (9) we first use the fact that *sb* and *ec* are parallel and conclude

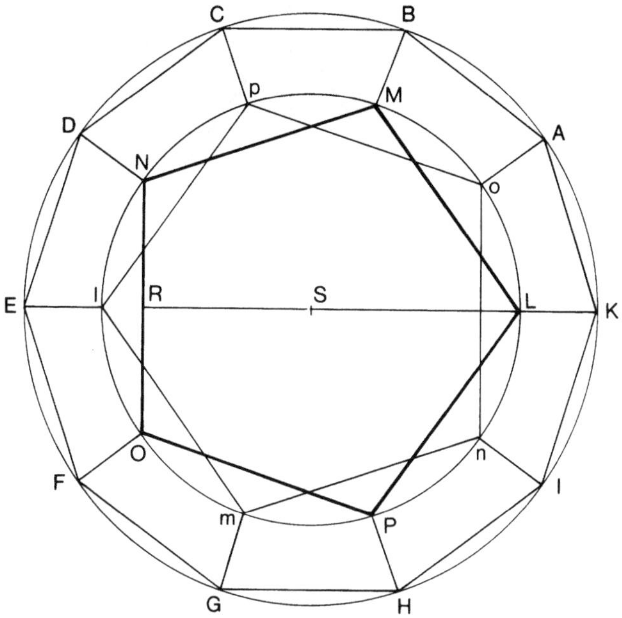

Figure 4. The lower part of Taylor's Figure 28, redrawn.

(since $vb = ba$) that $sb = \frac{1}{2}ev = \frac{1}{2}(\sigma + \delta)$. Moreover, from the similarity of the triangles hcg and hbs we get (cf. (7) and (2))

$$gc : sb = hc : hb = \sigma : \delta = \delta : (\sigma + \delta);$$

since $sb = \frac{1}{2}(\sigma + \delta)$ this relation shows that $gc = \frac{1}{2}\delta$, and hence $gv = gc$. Conversely, the point h can be obtained as the point of intersection of cb and sg when g is determined by (9). This was how Taylor determined h, and (8) gave him the points i, k, and l.

The elevation is completed when the points m and n (Figure 3) are constructed. For this construction Taylor used the observation that the original of mn is parallel to η and its midpoint is projected on s, which means $ms = sn = \frac{1}{2}\sigma$; and furthermore, that mn is perpendicular to ad. (The construction would have been slightly less complicated if Taylor had also used the fact that each of the triplets h, n, d; c, m, k; l, m, b lie on a line.)

The plan is easier to make: a geometrical consideration shows that the vertices of the dodecahedron is projected into two regular decagons having the same center and their vertices lying on radii from this center (Figure 4). The only problem left is to determine the ratio between the radii of the circles circumscribing the two decagons. To solve this problem we return to the elevation, or rather to the part of it which has been drawn in Figure 5. In this E is the projection upon the plan of the point c (cf. Figure 4) I that of e, and S that of s. The ratio between the two radii is

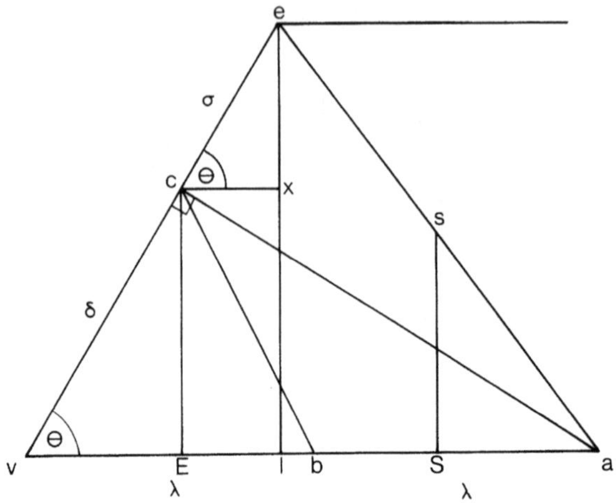

Figure 5. A part of the configuration of Figure 2.

equal to

$$ES : \mathbf{l}s.$$

Let x be the point of intersection of $e\mathbf{l}$ and the line through c parallel to va; the triangles ecx and avc are then similar hence

$$cx : ce = cv : va;$$

since $E\mathbf{l} = cx$, $ce = \sigma$, $cv = \delta$ and $va = 2\lambda$ this implies that (cf. (3))

$$E\mathbf{l} : \sigma = \delta : 2\lambda = \rho : \delta.$$

Using $\mathbf{l}S = \rho$ we get

$$E\mathbf{l} : \mathbf{l}S = \sigma : \delta,$$

from which it follows (cf. (2)) that

$$ES : \mathbf{l}S = (\sigma + \delta) : \delta = \delta : \sigma. \tag{10}$$

Since (cf. Figure 1) $\delta : \sigma = 2 \cos 36°$, the ratio $ES : \mathbf{l}S$ can be obtained from any construction leading to

$$ES : \mathbf{l}S = 2 \cos 36°,$$

and this is exactly what the construction described by Taylor did.

31. p. 102 (28), ll. 4–25: As in the previous example Taylor did not explain how he had derived this construction. Arguments similar to those used in Note 30 show that the construction is correct.
32. p. 102, (28), Cf. Errata, p. 72:
    ll. 15–20: For *O* read *R*.
    ll. 19–20: For *R* read *O*.
    ll. 19–20: For *T* read *X* and vice versa.

33. p. 105 (31), l. 11^-: For *RC* read *RB* (cf. Errata, p. 72).
34. p. 106 (32), l. 6^-: For *E* read *V*.
35. p. 107 (33), l. 11^-: *D* is supposed to be the point of intersection of *AE* and *BD*.
36. p. 112 (38), l. 1^-: For $a \times BC$ read $a \times AC$.
37. p. 118 (Plate 2), Figure 6: The points *P* and P^1 are interchanged.
38. p. 121 (Plate 5), Figure 16: The point *e* should have been to the other side of *V* because *BE* and *Oe* are supposed to be parallel.
39. p. 123 (Plate 7), Figure 21: There is a drawing mistake because *OF* is supposed to bisect the angle *EOD*.
40. p. 124 (Plate 8), Figure 24: A line segment *L* which should indicate the *distance* is missing.
41. p. 125 (Plate 9), Figure 25: As in the previous figure there is missing a line segment *L* to indicate the *distance*.

Book Three

New Principles of Linear Perspective

Brook Taylor

NEW

PRINCIPLES

OF

Linear Perſpective:

OR THE

ART of *DESIGNING*

ON A

PLANE

THE

REPRESENTATIONS

Of all ſorts of

OBJECTS,

In a more General and Simple METHOD than has been done before.

BY

Brook Taylor, LL.D. and R.S.S.

LONDON:

Printed for R. KNAPLOCK at the Biſhop's Head in St. *Paul's* Church-yard. MDCCXIX.

PREFACE.

Conſidering how few, and how ſimple the Principles are, upon which the whole Art of PERSPECTIVE *depends, and withal how uſeful, nay how abſolutely neceſſary this Art is to all ſorts of Deſigning; I have often wonder'd, that it has ſtill been left in ſo low a degree of Perfection, as it is found to be, in the Books that have been hitherto wrote upon it. Some of thoſe Books indeed are very voluminous: but then they are made ſo, only by long and tedious Diſcourſes, explaining of common things; or by a great number of Examples, which indeed do make ſome of theſe Books valuable, by the great Variety of curious Cuts that are in them; but do not at all instruct the Reader, by any Improvements made in the Art it self. For it seems that those, who have hitherto treated of this Subject, have been more conversant in the Practice of Designing, than in the Principles of Geometry; and therefore when, in their Practice, the Occasions that have offer'd, have put them upon inventing particular Expedients, they have thought them to be worth communicating to the Public, as Improvements in this Art; but they have not been able to produce any real Improvements in it, for want of a sufficient Fund of Geometry, that might have enabled them to render the Principles of it more universal, and more convenient for Practice. In this Book I have endeavour'd to do this; and have done my utmost to render the Principles of the Art as general, and as universal as may be, and to devise such Constructions, as might be the most simple and useful in Practice.*

In order to this, I found it absolutely necessary to consider this Subject entirely anew, as if it had never been treated of before; the Principles of the old Perspective being so narrow

narrow, and ſo confined, that they could be of no uſe in my Deſign: And I was forced to invent new Terms of Art, thoſe already in uſe being ſo peculiarly adapted to the imperfect Notions that have hitherto been had of this Art, that I could make no uſe of them in explaining thoſe general Principles I intended to eſtabliſh. The Term of Horizontal Line, for inſtance, is apt to confine the Notions of a Learner to the Plane of the Horizon, and to make him imagine, that that Plane enjoys ſome particular Privileges, which make the Figures in it more eaſy and more convenient to be deſcribed, by the means of that Horizontal Line, than the Figures in any other Plane; as if all other Planes might not as conveniently be handled, by finding other Lines of the ſame nature belonging to them. But in this Book I make no difference between the Plane of the Horizon, and any other Plane whatſoever; for ſince Planes, as Planes, are alike in Geometry, it is moſt proper to conſider them as ſo, and to explain their Properties in general, leaving the Artiſt himſelf to apply them in particular Caſes, as Occaſion requires.

My

My Design in this Book, is not to trouble the Reader with a Multitude of Examples, nor to descend to a perfect Explanation of any particular Cases; but to explain the general Principles of Perspective : *Which if I have been so happy as to have done, in such a manner as may be intelligible to the Reader; I don't doubt but he will easily be inclined to pardon my Shortness: For he will find much more pleasure in observing how extensive these Principles are, by applying them to particular Cases which he himself shall devise, while he exercises himself in this Art, than he would do in reading the tedious Explanations of Examples devised by another.*

I find that many People object to the first Edition that I gave of these Principles, in the little Book entituled, Linear Perspective, &c. *because they see no Examples in it, no curious Descriptions of Figures, which other Books of* Perspective *are commonly so full of; and seeing nothing in it but simple Geometrical Schemes, they apprehend it to be dry and unentertaining, and so are loth to give themselves the trouble to read it. To satisfy*

ſatisfy theſe nice Perſons in ſome meaſure, I have made the Schemes in this Book ſomething more ornamental, that they may have ſome viſible Proofs of the vaſt Advantages theſe Principles have over the common Rules of Perſpective, *by ſeeing what ſimple Conſtructions, and how few Lines are neceſſary to deſcribe ſeveral Subjects, which in the common Method would require an infinite Labour, and a vaſt Confuſion of Lines. It would have been eaſy to have multiplied Examples, and to have enlarged upon ſeveral things that I have only given Hints of, which may eaſily be purſued by thoſe who have made themſelves Maſters of theſe Principles. Perhaps ſome People would have been better pleaſed with my Book, if I had done this: but I muſt take the freedom to tell them, that tho' it might have amuſed their Fancy ſomething more by this means, it would not have been more inſtructive to them: for the true and beſt way of learning any Art, is not to ſee a great many Examples done by another Perſon; but to poſſeſs ones ſelf firſt of the Principles of it, and then to make them familiar, by exerciſing ones ſelf in the*

Practice

Practice. For it is Practice alone, that makes a Man perfect in any thing.

The Reader, who underſtands nothing of the Elements of Geometry, can hardly hope to be much the better for this Book, if he reads it without the Aſſiſtance of a Maſter; but I have endeavour'd to make every thing ſo plain, that a very little Skill in Geometry may be ſufficient to enable one to read this Book by himſelf. And upon this occaſion I would adviſe all my Readers, who deſire to make themſelves Maſters of this Subject, not to be contented with the Schemes they find here; but upon every Occaſion to draw new ones of their own, in all the Variety of Circumſtances they can think of. This will take up a little more Time at firſt; but in a little while they will find the vaſt Benefit of it, by the extenſive Notions it will give them of the Nature of theſe Principles.

The Art of Perſpective *is neceſſary to all Arts, where there is any occaſion for Deſigning;* *as* Architecture, Fortification, Carving, *and generally all the Mechanical Arts; but it is more particularly neceſſary to the Art of* Painting, *which can do nothing with-*

without it. A Figure in a Picture, which is not drawn according to the Rules of Perspective, *does not represent what is intended, but something else. So that it seems to me, that a Picture which is faulty in this particular, is as blameable, or more so, than any Composition in Writing, which is faulty in point of Orthography, or Grammar. It is generally thought very ridiculous to pretend to write an Heroic Poem, or a fine Discourse upon any Subject, without understanding the Propriety of the Language wrote in; and to me it seems no less ridiculous for one to pretend to make a good Picture without understanding* Perspective: *Yet how many Pictures are there to be seen, that are highly valuable in other respects, and yet are entirely faulty in this point? Indeed this Fault is so very general, that I cannot remember that I ever have seen a Picture, that has been entirely without it; and what is the more to be lamented, the Greatest Masters have been the most guilty of it. Those Examples make it to be the less regarded; but the Fault is not the less, but the more to be lamented, and deserves the more*

Care in preventing it for the future. The great Occasion of this Fault, is certainly the wrong Method that generally is used in the Education of Persons to this Art: For the Young People are generally put immediately to Drawing, and when they have acquired a Facility in that, they are put to Colouring. And these things they learn by rote, and by Practice only; but are not at all instructed in any Rules of Art. By which means when they come to make any Designs of their own, tho' they are very expert at drawing out, and colouring every thing that offers it self to their Fancy; yet for want of being instructed in the strict Rules of Art, they don't know how to govern their Inventions with Judgment, and become guilty of so many gross Mistakes, which prevent themselves, as well as others, from finding that Satisfaction, they otherwise would do, in their Performances. To correct this for the future, I would recommend it to the Masters of the Art of Painting, *to consider if it would not be necessary to establish a better Method for the Education of their Scholars, and to begin their Instructions with the Technical*

Parts

Parts of Painting, *before they let them loose to follow the Inventions of their own uncultivated Imaginations.*

The Art of Painting, *taken in its full Extent, consists of two Parts; the Inventive, and the Executive. The Inventive Part is common with* Poetry, *and belongs more properly and immediately to the Original Design (which it invents and disposes in the most proper and agreeable manner) than to the Picture, which is only a Copy of that Design already formed in the Imagination of the Artist. The Perfection of this Art of* Painting *depends upon the thorough Knowledge the Artist has of all the Parts of his Subject; and the Beauty of it consists in the happy Choice and Disposition that he makes of it: And it is in this that the Genius of the Artist discovers and shews it self, while he indulges and humours his Fancy, which here is not confined. But the other, the Executive Part of* Painting, *is wholly confined, and strictly tied to the Rules of Art, which cannot be dispensed with upon any account; and therefore in this the Artist ought to govern himself intirely by the Rules of Art, not to take any liberties whatſoever. For any thing that is not truly drawn according to the Rules of* Perſpective, *or not truly colour'd, or truly ſhaded, does not appear to be what the Artiſt intended, but ſomething elſe. Wherefore if at any time the Artiſt happens to imagine, that his Picture would look the better, if he ſhould ſwerve a little from theſe Rules, he may aſſure himſelf, that the Fault belongs to his Original Deſign, and not to the Strictneſs of the Rules; for what is perfectly agreeable and juſt in the real Original Objects themſelves, can never appear defective in a Picture, where thoſe Objects are exactly copied.*

Therefore to offer a ſhort Hint of the Thoughts I have ſometime had upon the Method which ought to be follow'd in inſtructing a Scholar in the Executive Part of Painting; *I would firſt have him learn the moſt common Effections of Practical Geometry, and the firſt Elements of Plain Geometry, and common Arithmetic. When he is ſufficiently perfect in theſe, I would have him learn* Perſpective. *And when he has made ſome progreſs in this, ſo as to have prepared his*

Judg-

Judgment with the right Notions of the Alterations that Figures muſt undergo, when they come to be drawn on a Flat, he may then be put to Drawing by View, and be exerciſed in this along with Perſpective, *till he comes to be ſufficiently perfect in both. Nothing ought to be more familiar to a* Painter *than* Perſpective; *for it is the only thing that can make the Judgment correct, and will help the Fancy to invent with ten times the eaſe that it could do without it. For the Colouring; before the Young Artiſt is employ'd in copying of Pictures, where there are great Variety of Colours to be imitated, it would be well that he ſhould be inſtructed in the Theory of the Colours; that he ſhould learn to know their particular Properties, their different Relations, and the various Effects that are produced by their Mixture; and that he ſhould be made well acquainted with the Nature of the ſeveral material Colours that are uſed in* Painting. *Theſe things ought to be learnt in a regular Method; and the Artiſt ought not to depend intirely upon the ſeveral indigeſted Obſervations, that may occur to him in Practice. And to this*

laſt

laſt Purpoſe of Colouring, I cannot help thinking, that the Theory I have endeavour'd to explain in the Appendix, *from Sir* Iſaac Newton, *may be of very great uſe to Learners. There may be regular Methods alſo invented for teaching the Doctrine of Light and Shadow; and other Particulars relating to the Practical Part of* Painting, *may be improved and digeſted into proper Methods for inſtructing the Young Artiſts. But I only hint at theſe things, recommending them to the Maſters of the Art to reflect and improve upon.*

The Book it ſelf is ſo ſhort, that I need not detain the Reader any longer in the Preface, *by giving him a more particular Account of what he may expect to find in it.*

LINEAR

LINEAR PERSPECTIVE.

PART I.
DEFINITIONS.

DEFIN. I.

LINEAR PERSPECTIVE is the Art of describing exactly, on a Plane Surface, the Representations of any given Objects.

TO have a compleat and clear Notion of the Principles of this Art, let the Reader consider, that a Picture drawn in the utmost degree of Perfection, and placed in a proper Position, ought so to appear to the Spectator, that he should not be able to distinguish what is there represented, from the real original Objects actually placed

placed where they are repreſented to be. In order to produce this Effect, it is neceſſary that the Rays of Light ought to come from the ſeveral Parts of the Picture to the Spectator's Eye, with all the ſame Circumſtances of Direction; Strength of Light and Shadow, and Colour, as they would do from the correſponding Parts of the real Objects ſeen in their proper Places. Thus (*Fig.* 1.) ſuppoſing the Spectator, whoſe Eye is O, to be looking on the Picture of a Cube *abcde*, and ABCDE to be the real original Cube, actually placed where it ſeems to him to be; the Light from any Point *a* of the Picture ought to come to the Spectator's Eye O by the Ray *a*O, in the ſame Direction, with the ſame Colour, and with the ſame Strength of Light and Shadow, as it would do from the correſponding Point A of the Original Cube, by the Ray AO. The three Circumſtances juſt mention'd make the executive Part of the Art of Painting to conſiſt of three Parts, *viz.* Drawing, (which relates wholly to the Poſition, and conſequently to the Shapes of the Figures on the Picture; and which when it is done exactly by Mathematical Rules, and not by an acquired Habit of the Hand and Eye, is *Perſpective*, the Subject of the preſent Diſcourſe;) Colouring, and the Art of Light and Shadow, which the *Italians* call the *Chiaroſcuro*. The Principles of all theſe Parts of Painting are to be drawn from this general Conſideration, and particularly thoſe of *Perſpective*. Wherefore in the Demonſtrations of the following Propoſitions in this Book, we muſt always have recourſe to this general Foundation, by ſhewing that the Rays of Light will come

come in the ſame Directions from the ſeveral Parts aſſigned in the Picture, as they would do if they came from the correſponding Parts of the original Objects placed in their proper Situations.

DEFIN. II.

When Lines drawn according to a certain Law from the ſeveral Parts of any Figure, cut a Plane, and by that Cutting or Interſection deſcribe a Figure on that Plane, that Figure ſo deſcribed is called the *Projection* of the other Figure. The Lines producing that Projection, taken all together, are called the *Syſtem of Rays*. And when thoſe Rays all paſs thro' one and the ſame Point, they are called the *Cone of Rays*. And when that Point is conſider'd as the Eye of a Spectator, that Syſtem of Rays is called the *Optic Cone*.

DEFIN. III.

When the Syſtem of Rays are all parallel to each other, and perpendicular to the Horizon, and the Projection is made on a Plane parallel to the Horizon, it is called the *Ichnography* of the Figure pro-

C poſed.

poſed. Thus (*Fig.* 2.) the Plane GHIK being parallel to the Horizon, and the Rays A*a*, B*b*, C*c*, *&c.* coming from the ſeveral Parts of the *Octaedron* ABC DEF, being perpendicular to it, and parallel to each other, the Projection *a b c de* made by them is the *Ichnography* of the Figure ABCDEF.

DEFIN. IV.

When the Syſtem of Rays are parallel to each other, and to the Horizon, and the Projection is made on a Plane perpendicular to thoſe Rays, and to the Horizon, it is called the *Orthography* of the Figure propoſed. Thus (*Fig.* 2.) the Rays A a, B b, C c, *&c.* being parallel to the Horizon, and to each other, and the Plane GHLM, on which the Projection is made, being perpendicular to them, a b c d e f is the *Orthography* of the Figure ABCDEF.

Theſe are the common Definitions of the Terms *Ichnography* and *Orthography*, but we ſhall hereafter uſe them to ſignify any two Projections that are made by Syſtems of Parallel Rays, when thoſe Syſtems are perpendicular to each

each other, and to the Planes on which the Projections are made, as in the present Figure, without having any regard to their Situation with respect to the Horizon.

In this kind of Projections, the Projection of any particular Point, or Line, is sometimes called the Seat of that Point, or Line, on the Plane of the Projection. Thus *a* is the Seat of the Point A on the Plane G H I K, and a f is the Seat of the Line A F on the Plane G H L M.

DEFIN. V.

When the Projection is made by a Cone of Rays, it is called the *Schenography*. Thus *(Fig.* 1.*)* the Figure *a b c d e* projected on the Plane F G H I by the Rays A O, B O, C O, *&c.* coming from the several Parts of the Cube A B C D E to the Point O, is the *Schenography* of the Figure A B C D E.

We shall hereafter shew this Projection to be the Picture of the Object A B C D E, to be seen by a Spectator's Eye at O.

It is alsо evident that the Shadows of Figures are this Projection, when the Light is consider'd as a single Point. Tho' in the Case of the Sun or Moon, that Point being at an infinite Distance (as to all Sense) the Projecting Rays are parallel to each other.

DEFIN. VI.

The *Point of Sight* is that Point, where the Spectator's Eye ought to be placed, to look upon the Picture.

This Point is no other than the Vertex of the Optic Cone, as will be evident from *Theor.* 2. where we ſhew, that the Repreſentation of any Object is no other than its *Schenographic* Projection on the Plane of the Picture.

DEFIN. VII.

If from the Point of Sight there be drawn a Line perpendicular to the Picture, the Point where that Line cuts the Picture is called the *Center* of the Picture. And the Diſtance between that Center and the Point of Sight, is called the Diſtance of the Picture.

DEFIN. VIII.

If thro' the Point of Sight there be imagined to paſs a Plane parallel to the Picture, that Plane is called the *Directing Plane*.

DEFIN.

DEFIN. IX.

By *Original Object* (whether it be Point, Line, Surface or Solid) we mean the real Object placed in the Situation it is repreſented to have by the Picture.

DEFIN. X.

By *Original Plane*, we mean the Plane wherein is ſituated any Original Point, Line, or Plane Figure.

DEFIN. XI.

The Point where any Original Line (continued if need be) cuts the Picture, is called ſimply the *Interſection* of that Line.

DEFIN. XII.

The Line wherein any Original Plane cuts the Picture, is called ſimply the *Interſection* of that Original Plane.

DEFIN. XIII.

The Point where any Original Line cuts the Directing Plane is called the *Direct-*

ing

ing Point of that Original Line. And a Line drawn thro' that Directing Point, and the Point of Sight, is call'd the *Director* of that Original Line.

DEFIN. XIV.

The Line wherein any Original Plane cuts the Directing Plane, is call'd the *Directing Line* of that Original Plane.

DEFIN. XV.

A Line drawn thro' the Point of Sight parallel to any Original Line, is called ſimply the *Parallel* of that Original Line.

DEFIN. XVI.

A Plane paſſing thro' the Point of Sight parallel to any Original Plane, is called ſimply the *Parallel* of that Original Plane.

DEFIN. XVII.

The Point where the Parallel of any Original Line cuts the Picture, is call'd the *Vaniſhing Point* of that Line. And the Diſtance

Diſtance between that Vaniſhing Point and the Point of Sight, is called ſimply the *Diſtance* of that Vaniſhing Point.

DEFIN. XVIII.

The Line wherein the Parallel of any Original Plane cuts the Picture, is called the *Vaniſhing Line* of that Plane. And if from the Point of Sight there be drawn a Line cutting that Vaniſhing Line at Right Angles, the Point where that Vaniſhing Line is ſo cut, is called the Center of it. And the Diſtance between that Center and the Point of Sight, is called ſimply the *Diſtance* of that Vaniſhing Line.

DEFIN. XIX.

The Repreſentation of any Figure is called the *Projection* of that Figure.

In order to comprehend the Senſe of theſe Definitions more fully, let the Reader imagine the Plane ABC (*Fig.* 3.) to be the Surface of the Picture, O to be the Point of Sight (*Def.* 6.) the Plane ODE to be the Directing Plane parallel to the Picture (*Def.* 8.) FG to be an Original Line (*Def.* 9.) in the Original Plane FGH (*Def.* 10.) cutting

cutting the Picture in B I, and the Directing Plane in D E. And let O A C be a Plane parallel to the Original Plane F G H, and cutting the Picture in the Line A C; and let the Line O V be parallel to the Original Line F G, and cut the Picture in V. And let F G cut the Picture in B, and the Directing Plane in D. Theſe things being ſuppoſed, B is the Interſection of the Original Line F G (*Def.* 11.) D its Directing Point (*Def.* 13.) and V its Vaniſhing Point (*Def.* 17.) and O D its Director (*Def.* 13.) and O V is the Diſtance of the Vaniſhing Point V (*Def.* 17.) B I is the Interſection (*Def.* 12.) D E the Directing Line (*Def.* 14.) O A C the Parallel (*Def.* 16.) and A C the Vaniſhing Line (*Def.* 18.) of the Original Plane F G H. And if there be drawn O S cutting the Vaniſhing Line A C at Right Angles in S, that Point S will be the Center (*ib.*) and S O will be the Diſtance (*ib.*) of the Vaniſhing Line A C.

AXIOM I.

The common Interſection of two Planes is a ſtraight Line.

AXIOM II.

If two ſtraight Lines meet in a Point, or are parallel to one another, there may be a Plane paſſing thro' them both.

AXIOM III.

If three ſtraight Lines cut one another, or if two of them being parallel, are both cut by the third, they will all three be in the ſame Plane; that is, a Plane paſſing thro' any two of them will alſo paſs thro' the third.

AXIOM IV.

Every Point in any ſtraight Line, is in any Plane that Line is in.

LEMMA I.

If BSO (Fig. 4.) *and* AEBD *be two Planes, cutting each other in the Line* ASB, *and from any Point* O *of one of the Planes be drawn two Lines* OS *and* OC *cutting the Line* AB, *and the other Plane* AEBD *at Right Angles in* S *and* C, *and there de drawn* CS; *that Line* CS *will be perpendicular to* ASB.

This follows from *Prop.* 11. *Lib.* 11. *Elem.*

THEOREM I.

A Line drawn from the Center of the Picture to the Center of a Vaniſhing Line, is perpendicular to that Vaniſhing Line.

DEMONSTRATION.

Imagine AEBD (*Fig.* 4.) to be the Plane of the Picture, O to be the Point of Sight, and OSB to be a Parallel Plane producing the Vaniſhing Line ASB by its Interſection with the Picture; and let S be the Center of that Vaniſhing Line, and C be the Center of the Picture. Then having

D drawn

drawn OS and OC, draw CS. OS is perpendicular to AS (by *Def.* 18.) and OC is perpendicular to the Plane AEBD (by *Def.* 7.) Therefore CS is perpendicular to AS (by *Lem.* 1.) Which was to be proved.

COROL. The Diſtance OS of any Vaniſhing Line ASB, is the Hypothenuſe of a Right-angled Triangle, whoſe Legs are the Diſtance of the Picture OC, and the Diſtance CS between the Center of that Vaniſhing Line and the Center of the Picture.

THEOR. II.

* *The Perſpective Repreſentation, or Projection, of any Object, is the ſame as the Ichnographic Projection of it on the Plane of the Picture, the Point of Sight being the Vertex of the Optic Cone.*

DEMONSTRATION.

For by the Explanation of the Principles of the Art of Painting, in *Def.* 1. ſince the Light muſt come to the Spectator's Eye O (*Fig.* 1.) in the ſame Direction from any Point *a* of the Projection, as it would do from the correſponding Point A of the original Object, it is evident that the Rays *aO* and AO are in one and the ſame ſtraight Line. Whence it is evident, that the Projection *a* is the Interſection of the Picture with the Ray AO, and the whole Projection *abcde* is the Schenographic Projection of the original Figure ABCDE made by the Optic Cone OABCDE, whoſe

whoſe Vertex is the Point of Sight O. Which was to be proved.

COROL. 1. The Projection of a ſtraight Line is a ſtraight Line. For the Optic Cone OAB, which produces the Projection *de* of any Line DE, is the Surface of a Plane Triangle ODE, all the Rays going to O from the ſeveral Points of the Line DE being in the Plane paſſing thro' the Lines DO and EO. Therefore *de* is the Interſection of the Picture with the Plane Triangle ODE, and conſequently is a ſtraight Line (by *Ax.* 1.)

COROL. 2. The Original of a Projection may be any Object, that will produce the ſame Cone of Rays. Thus the Original of the Projection *de*, may be any Line d e, which produces the Optic Cone ODE, as well as the Line DE.

This being ſo, it may reaſonably be ask'd, whence it comes that Figures drawn on a Picture appear to be what they are deſigned to repreſent. The Reaſon is, becauſe the Mind has got a Habit of judging Objects, that are ſo and ſo related, have ſuch and ſuch Colours, and are ſo and ſo enlightned and ſhaded, to be of ſuch and ſuch a Shape, and to be ſo and ſo ſituated. Theſe Circumſtances are all of them neceſſary to make a Picture compleat, tho' the ſimple Drawing is ſometimes almoſt ſufficient, upon account of the Relation of the Parts; as in a Pavement, where all the Stones appear to be ſquare, tho' they are repreſented by very irregular Figures. I ſay, it is the Relation of the Parts which produces this Effect; for the Repreſentation of any one of the ſingle Squares would hardly appear to be ſquare, if there were no other Objects to biaſs the Judgment by their Relation to it.

THEOR. III.

The Projection of a ſtraight Line not parallel to the Picture, paſſes thro' both its Interſection and Vaniſhing Point.

DEMONSTRATION.

All other things in *Fig.* 3. being underſtood (as above at the End of the Definitions) let *fg* be the Projection of the Original Line FG, FO, and GO being the Rays which produce the Projections *f* and *g* of the Points F and G (See *Th.*2.) By *Theor.* 2. *fg* is the Interſection of the Picture with the Plane of the Triangle OFG. But all the Line FGB is in that Plane; and conſequently the Interſection B. Wherefore *fg* continued muſt paſs thro' B. Which was firſt to be proved.

OV being parallel to FG (*Def.* 17.) is in the ſame Plane of the Triangle OFG. Therefore *fg* continued will paſs thro' V. Which was alſo to be proved.

This Theorem being the principal Foundation of all the Practice of *Perſpective*, the Reader would do well to make it very familiar to him. To help him a little in his Reflections upon it, I have again repreſented the Senſe of it in *Fig.* 1. where the Projection *bc* meets the Original Line BC in its Interſection K, and paſſes alſo thro' its Vaniſhing Point V, which is produced by its Parallel OV.

N.B.

N.B. When the Original Line it ſelf paſſes through its Vaniſhing Point, the whole Projection of it will be that Point; ſo that in that caſe the Line may be ſaid to vaniſh. This is one Reaſon for my uſing that Term. Another Reaſon is, that the further any Object is off, upon any Line, the ſmaller is its Projection, and at the ſame time, the nearer to this Point; and when it comes into this Point, its Magnitude vaniſhes, becauſe the Original Object is at an infinite Diſtance. This is eaſily conceived by imagining a Man to be going from you in a long Walk, who appears to be ſmaller and ſmaller, the further he goes. The Reaſon of this Diminution will appear from the following *Corollaries*.

COROL. 1. The Projections of all Original Lines that are parallel to one another, but not to the Picture, paſs thro' the ſame Vaniſhing Point: For they have but one Parallel common to them all, and conſequently but one Vaniſhing Point. This may be ſeen repreſented in *Fig.* 1. where the Projection *da* and *cb* of the Parallel Lines DA and CB meet in their common Vaniſhing Point V.

COROL. 2. The Center of the Picture is the Vaniſhing Point of Lines perpendicular to the Picture. (See *Defin.* 7, 15, & 17.)

THEOR. IV.

The Projection of a Line parallel to the Picture, is parallel to its Original.

DEMON-

DEMONSTRATION.

Let the Plane EF (*Fig.* 5.) be the Picture, AB be the Original Line parallel to it, and *ab* its Projection; O being the Point of Sight, and OAB the Optic Cone. By *Theor.* 2. *ab* is the Interfection of the Picture with the Plane of the Triangle OAB. Therefore AB being parallel to the Plane EF, *ab* is parallel to AB. For they are both in the Plane of the Triangle OAB, and do not meet; for if they did, their common Interfection would be in the Plane EF, and confequently AB would not be parallel to the Plane EF.

COROL. 1. The Projections of feveral Lines parallel to one another and to the Picture, are parallel to one another. Thus *ab* and *dc* are parallel to each other, and to their Originals AB and CD.

COROL. 2. The Projection *abcd* of any Plane Figure ABCD parallel to the Picture, is fimilar to its Original. For having drawn the Diagonal AC, and its correfponding Projection *ac*, the Sides *ab*, *bc*, *ac*, are parallel to their correfponding Originals AB, BC, AC; wherefore the Angles at *a*, *b*, and *c*, are equal to the correfponding Angles at A, B, and C; and confequently the Triangle *abc* is fimilar to the Triangle ABC. For the fame reafon *acd* is fimilar to ACD; and confequently the Figure *abcd* is fimilar to ABCD.

COROL. 3. In the fame cafe, the Length of any Line *ab* in the Projection, is to the Length of

2

of its Original AB, as the Diſtance of the Picture is to the Diſtance between the Point of Sight and the Plane of the Original Figure. Let O*g*G be perpendicular to thoſe two Planes, cutting them in *g* and G. Then will ***ab*** : AB :: O***a*** ; OA :: O*g* : OG (by *Prop.* 17. *Lib.* 11. *Elem.*) But *g* is the Center, O*g* is the Diſtance of the Picture, and OG is the Diſtance between the Point of Sight O, and the Original Plane ABCD. Wherefore ***ab*** is to AB, as the Diſtance of the Picture is to the Diſtance between the Point of Sight and the Plane of the Original Figure.

THEOR. V.

The Projection of a Line is parallel to its Director.

In *Fig.* 3. as already explain'd at the End of the Definitions, and in *Theor.* 3. the Lines OF, OG, OD, *fg*, are all in the ſame Plane. But the Directing Plane ODE is parallel to the Plane of the Picture ABIC (*Def.* 8.) Therefore the Director OD is parallel to the Projection *fg* (by *Prop.* 16. *Lib.* 11. *Elem.*)

COROL. 1. The Projections of Lines that have the ſame Director, are parallel to each other.

COROL. 2. When the Original Line is parallel to the Picture, its Director is parallel to it, and conſequently is in the Parallel of any Plane paſſing thro' that Original Line; and therefore the Vaniſhing Line of that Plane, and the Projection of the Line, are parallel to one another. *

THEOR.

THEOR. VI.

The Vanishing Line, Intersection, and Directing Line of any Original Plane, are parallel to each other.

In the same *Fig.* understood as has been already explain'd, the Planes OVC and DFH being parallel (*Def.* 16.) as also ODE and CAB (*Def.* 8.) the Vanishing Line CV, the Intersection IB, and the Directing Line ED, are parallel to each other (by *Prop.* 16. *Lib.* 11. *Elem.*) Which was to be proved.

COROL. The Distance *f*V between the Projection of any Point, and the Vanishing Point V, is to the Distance BV, between the Intersection B, and the same Vanishing Point V, as the Distance OV of that Vanishing Point V, is to the Distance DF, between the Director and the Original Point F. For OVBD is a parallelogram, wherefore BV is equal to DO, and the Triangles *f*OV and OFD are similar, because their Sides V*f* and OD are parallel. Wherefore *f*V : VO :: DO (=BV): DF.

THEOR. VII.

The Vanishing Points of all Lines in any Original Plane, are in the Vanishing Line of that Plane.

DEMON-

DEMONSTRATION.

For ſince all the Original Lines are in the ſame Plane, their Parallels, which all paſs thro' the Point of Sight, will be all of them in the Parallel Plane (by *Prop.* 15. *Lib.* 11. *Elem.*) wherefore all the Vaniſhing Points are in the Vaniſhing Line.

COROL. 1. Original Planes, that are parallel, have the ſame Vaniſhing Line.

COROL. 2. The Vaniſhing Point of the common Interſection of two Original Planes, is the Interſection of their Vaniſhing Lines.

COROL. 3. The Vaniſhing Line of a Plane perpendicular to the Picture, paſſes through the Center of the Picture.

THEOR. VIII.

The Interſections of all Lines in the ſame Original Plane, are in the Interſection of that Plane.

This needs no *Demonſtration.*

COROL. 1. The Interſection of the common Interſection of two Original Planes, is the Interſection of their Interſections.

COROL. 2. Planes, whoſe common Interſection is parallel to the Picture, have parallel Interſections, and alſo parallel Vaniſhing Lines.

E PRO-

* PROBLEM I.

Having given the Center and Diſtance of the Picture, to find the Projection of a Point, whoſe Seat on the Picture, with its Diſtance from it, are given.

Let S (*Fig.* 6.) be the Center of the Picture, and *b* the given Seat of the Original Point. Draw at pleaſure SO equal to the Diſtance of the Picture, and parallel to it draw *b*A equal to the Diſtance of the Original Point from its Seat. Draw S*b* and AO meeting in *a*, which will be the Projection ſought.

DEMONSTRATION.

If the Angle OS*b* (and conſequently A*ba*) had been a Right Angle; then turning the Triangles SO*a* and *b*A*a* round the Line S*ab* as an Axis, till SO and *b*A become perpendicular to the Picture, O would be the Point of Sight, and A the Original Point, and AO would be the Viſual Ray cutting the Picture in *a*, which conſequently would be the Projection of the Point A, by *Theor.* 2. But the Point *a* is the ſame, whether the Angle OS*b* be a right Angle or not, becauſe the Triangles OS*a* and A*ba* are ſimilar, and therefore S*a* : *ab* : : SO : *b*A, which Proportionality is not affected by altering the Angle OS*b*. Therefore in all caſes the Point *a* thus found, is the Projection ſought of a Point, whoſe Seat

Seat on the Picture is *b*, and its Distance from its Seat is *b* A.

COROL. 1. Having drawn S *b*, the Point *a* may be found by a Scale and Compasses dividing the Line S *b* in *a*, so that S *a* may be to *a b*, as the Distance of the Picture SO, is to the Distance *b* A of the Original Point from its Seat *b*.

COROL. 2. By this Proposition the Projection of any Line may be found, by finding the Projections of two Points in it, and drawing a Line thro' those two Points Projections.

PROBLEM II.

To find the Projection of a Line, its Vanishing Point and Distance, having given its Seat, Intersection, and the Angle it makes with its Seat, and the Center and Distance of the Picture.

Let DE *Fig.* 6. be the given Seat of the Line proposed, D its Intersection, and S the Center of the Picture. Draw D C making the Angle EDC equal to the Angle the Original Line makes with its Seat. Draw S V parallel to D E, and SO perpendicular to it, and equal to the Distance of the Picture. Then draw O V parallel to D C cutting S V in V, and draw D V. Then will V be the Vanishing Point and O V its Distance, and DV the indefinite Representation of the Line proposed.

DEMONSTRATION.

Imagine the Planes OSV and CED to be turned round the Lines SV and DE as Axes, till they become perpendicular to the Picture. Then will O be the Point of Sight, and DC the Original Line; and OV being parallel to it, V is its Vaniſhing Point (by *Def.* 17.) and conſequently DV is its Projection (by *Theor.* 3.) Which was to be proved.

COROL. 1. DAC being conceived as the Original Line laid on the Picture, by turning the Plane CDE round the Line ED, the Projection of any part of it AC may be found by drawing the Lines AO and CO as Viſual Rays, cutting DV in *a* and *c*. For the Points *a* and *c* depend only on the Paralleliſm of the Lines OV and DC, and their proportion; *a*V being to *a*D as VO is to DA, and *c*V : *c*D : : VO : DC, upon account of the ſimilar Triangles *a*VO and *a*DA, and *c*VO and *c*DC. To underſtand this more clearly, the Reader may compare this Figure with *Fig.* 3. where the Points O, V, *f*, *g*, B, G, F, are analogous to the Points O, V, *c*, *a*, D, A, C, reſpectively.

COROL. 2. Having found DV, the Projection *c* of any Point C may be found by a Scale and Compaſſes, as making *c* V : *c* D : : OV : CD.

PRO-

PROBLEM III. *

Having given the Projection of a Line, and its Vanishing Point; to find the Projection of the Point that divides the Original Line in any given Proportion.

Let AB (*Fig.* 7.) be the given Projection of the Line to be divided, and V its Vanishing Point. Draw at pleasure VO and *ba* parallel to it, and thro' any Point O of the Line VO draw OA and OB cutting *ba* in *a* and *b*. Divide *ab* in *c* in the Proportion given, and draw O*c* cutting AB in C. Then will C be the Projection sought, the Original of BC being to the Original of CA, as *bc* is to *ca*.

DEMONSTRATION.

OV being parallel to *ba*, *ba* may be consider'd as the Original Line and OV as its Parallel, and consequently O as the Point of Sight, and *a*O, *b*O, *c*O, as Visual Rays projecting the Points A, B, C.

COROL. The Mathematical Reader will easily find, that CA × BV : CB × AV :: *ca* : *cb*. Whence the Point C may be found by a Scale and Compasses, making CA : CB :: AV × *ca* : BV × *cb*.

PRO-

* PROBLEM IV.

Having given the Projection of a Line, and its Vaniſhing Point; from a given Point in that Projection, to cut off a Segment, that ſhall be the Projection of a given Part of the Original of the Projection given.

Let AB (*Fig.* 8.) be the Projection given, V its Vaniſhing Point, and C the Point from whence is to be cut off the Segment. Draw at pleaſure VO and *abc* parallel to it, and from any Point O in VO draw OA, OB, OC cutting *ab* in *a*, *b*, *c*. Make *cd* to *ab*, as the given Part is to the Original of AB, and draw O*d* cutting AB in D. Then will CD be the Segment ſought.

DEMONSTRATION.

If OV be conceived as the Vaniſhing Line of any Plane paſſing through the Original of ABV; *abcd* being parallel to it, may be conceived as the Projection of a Line parallel to the Picture (by *Cor.* 2. *Theor.* 5.) and therefore its Parts *ab* and *cd* will be in the ſame proportion to one another as their Originals (by *Theor.* 4.) But becauſe of the Vaniſhing Point O, the Originals of O*a*, O*b*, O*c*, O*d* are parallel (by *Cor.* 1. *Th.* 3.) Wherefore the Original of CD is to the Original of AB, as *cd* is to *ab* (by *Prop.* 2. *Lib.* 6. *Elem.*) Which was to be proved.

N.B. This Propoſition might have been demonſtrated as the foregoing, and the foregoing may

may be consider'd as a particular Case of this, *viz.* when the Point C of this Proposition coincides with one of the Points A or B.

COROL. The Point D may be found by a Scale and Compasses, making DC : DV :: *dc* × AB × CV : *ab* × AV × BV.

PROBLEM V. *

Having given the Center and Distance of the Picture; to find the Vanishing Line (with its Center and Distance) of a Plane, whose Intersection is given, with the Angle of its Inclination to the Picture.

Let AB (*Fig.* 9.) be the given Intersection of the Plane, and C the Center of the Picture. Draw CO parallel to AB, and equal to the Distance of the Picture, and draw CA perpendicular to AB. Draw OS cutting AC in S, so that the Angle OSC may be equal to the Inclination of the original Plane to the Picture. Draw SD parallel to AB. Then will SD be the Vanishing Line sought, S its Center, and OS its Distance.

DEMONSTRATION.

Imagine the Triangle OSC to be raised up on the Picture, so that OC may be perpendicular to the Picture. In that case O will be the Point of Sight, and SD being parallel to AB, a Plane passing thro' the Line SD and the Point O will be the Parallel of any original Plane passing thro'

AB,

AB, and inclined to the Picture in the Angle OSC. Wherefore SD is the Vaniſhing Line ſought. And OS, in that ſuppoſition, being perpendicular to SD, S is the Center, and SO the Diſtance of the Vaniſhing Line SD. Which was to be proved.

N. B. Taking OC for Radius, CS is the Cotangent, and OS the Co-ſecant of the Inclination of the Original Plane to the Picture.

* PROBLEM VI.

Having given the Interſection of an Original Plane, with its Vaniſhing Line, its Center and Diſtance; to find the Projection of any Line in the Original Plane, having the Original Figures drawn out in their juſt Proportions.

Let DF (*Fig.* 16.) be the Interſection given, HG the Vaniſhing Line, and S its Center. Draw SO perpendicular to GH, and equal to the Diſtance of the Vaniſhing Line GH, and let the Space X be the Original Plane, ſeen on the Reverſe, as Objects appear in a Looking Glaſs; the Space Y being the Parallel Plane in the ſame manner folded down on the Picture; and let AB be the Original Line, whoſe Projection is ſought. Let AB cut the Interſection in D, and draw OG parallel to AB, cutting the Vaniſhing Line in G. Draw DG, which will be the indefinite Projection of AB. Thro' A and B draw at pleaſure AC and BC meeting in C, and in the ſame manner find their indefinite Projections FI, and EH cutting DG in *a* and *b*. Then will

will *ab* be the determinate Projection of AB, *a* being the Projection of the Extremity A, and *b* the Projection of the Extremity B.

Otherwise. Suppose KL to be the Original Line given. Having found its indefinite Projection QG, as before, draw OK and OL cutting it in *k* and *l*, which will be the Projections of the Extremities K and L.

Otherwise, by the Directors.

Let DF (*Fig.* 11.) be the Interfection given. *
And let the Original Plane be folded down on the Picture, so as to bring the Directing Line into the Place HI, the Distance between AF *
and HI being equal to the Distance of the Vanishing Line given. Let also O be the Point of Sight brought into the Picture at the same time, along with the Directing Plane HOI. To find the indefinite Projection of any Original Line AB, continue it till it cuts EF in F and HI in G; then draw OG, and F*a* drawn parallel to it will be the indefinite Projection sought. Then finding in the same manner the indefinite Projection E*d* of any other Line AD passing thro' A, by its Interfection with F*a* is got the Projection *a* of the Extremity A. And in the same manner is got the other Extremity *b*. Or those Extremities might be found by drawing Lines from A and from B to O, as in the foregoing Construction.

DEMONSTRATION.

Imagine the Figures to be folded, *Fig.* 10. in DF and HG, and *Fig.* 11. in EF and HI, till

F the

the Original Plane, its Parallel, and the Directing Plane, and along with them the Point of Sight O, come into their proper Places. Then you will find, that D in *Fig.* 10. and F in *Fig.* 11. will be the Interſection of AB, and G in *Fig.* 11. will be its Directing Point. But OG in *Fig.* 10. is ſtill parallel to AB, wherefore G is its Vaniſhing Point, and DG its indefinite Projection (by *Theor.* 3.) and F*a* in *Fig.* 11. is ſtill parallel to OG, which is the Director of AB; wherefore F*a* is the indefinite Projection of AB (by *Theor.* 5.) That *a* found by the interſection of FI with DG in *Fig.* 10. and of E*d* with F*a* in *Fig.* 11. is the Projection of the interſection of the Original Lines AB and AC in *Fig.* 10. and of AB and AD in *Fig.* 11. is obvious. The other Conſtruction by the Lines AO is the ſame as by the Lines AO and CO for finding the Points *a* and *c* in *Fig.* 6. as is explain'd in *Cor.* 1. *Prob.* 2.

N.B. 1. The Reader that is not uſed to Mathematical Subjects may help himſelf in conceiving the Manner here deſcribed of bringing the Original Plane, its Parallel, and the Directing Plane, with the Point of Sight, into the Plane of the Picture, by conceiving in *Fig.* 3. thoſe ſeveral Planes to be brought together by enlarging the Angles OVB and ODB, till the Planes lie flat on one another, and the Original Plane is ſeen on the back-ſide.

N.B. 2. In *Fig.* 11. the Projections *ad* and *kl* are parallel, their Originals having both of them the ſame Director OH, according to *Cor.* 1. *Theor.* 5. The ſame may be obſerved in *lm* and *cd*, which have the ſame Director OI.

PRO-

PROBLEM VII. *

Having given the ſame things as in the foregoing Problem, to find the Projection of any Figure in the Original Plane.

This is done by finding the Projections of the ſeveral Parts of the Figure given, by the foregoing Problem.

For example, the Projection *klmnp* (*Fig.* 10.) of the Pentagon KLMNP is found thus. Drawing OG, OH, OI, OV parallel to KL, LM, MN, KP reſpectively, the Points G, H, I, and V, are their Vaniſhing Points. And KL, LM, MN, being continued cut the Interſections in their Interſections Q,R,T. Whence drawing QG, RH, TI, are got the Projections *l*, *m* of the Points L and M, by their mutual interſections. Then drawing OK and ON, are got the Points *k* and *n*. Then drawing *k*V to the Vaniſhing Point V of KP, is got the indefinite Projection of KP. Laſtly drawing OP is got the Point *p*.

The Projections of Curvelined Figures are to be got by finding the Projections of ſeveral of their Points, and afterwards joining them neatly by hand. Thus in *Fig.* 13. *n.* 1. DE being the Interſection, and VF the Vaniſhing Line, and O the Point of Sight, and ABC an Original Circle, placed as in the foregoing Problem, the Projection *a* of any Point A may be found by drawing at pleaſure AD, and OV parallel to it; then drawing DV and OA meeting in the Point ſought *a*, according to the Conſtruction in *Problem* 6. D *

F 2 being

being the Interſection, and V the Vaniſhing Point of the Line AD. And the ſeveral Lines AD being drawn parallel to one another, the ſame Vaniſhing Point V may ſerve for them all.

Or as in *N.* 2. VF being the Directing Line brought into the Picture, as in *Fig.* 11. and the reſt remaining as before; drawing at pleaſure AD cutting DE and VF in D and V, then drawing OV and D*a* parallel to it, the Projection *a* is got by drawing OA cutting D*a* in *a*. And the ſame Point V being uſed for all the Points A, all the Lines D*a* will be parallel to one another, and to the ſame Line OV.

PROBLEM VIII.

To find the Projection of any Figure in a Plane parallel to the Picture.

The Projection being ſimilar to its Original (by *Cor.* 2. *Theor.* 4.) this is done by making an exact Copy of the Original Figure; making the Homologous Sides in the Proportion explain'd in *Cor.* 3. of the ſame *Theorem.*

PROBLEM IX.

Having given the Interſection of a Plane, and its Vaniſhing Line, with its Center and Diſtance; to find the Original of any Projection given on the Picture.

Every thing being diſpoſed in *Fig.* 10. as in *Problem* 6, and 7. let it be propoſed to find the Original

Original of the Figure *klmnp*. Having continued the Projections *kl*, *lm*; *mn* till they cut the Interſection and Vaniſhing Line in their Interſections Q, R, T, and their Vaniſhing Points G, H, I, and *kp* in its Vaniſhing Point V, draw OG, OH, OI, OV, and QK, RM, TN parallel to the three firſt reſpectively meeting in L and M, which will be the Originals of *l* and *m*. Draw O*k* and O*n*, which will cut QL and TM in the Originals K and N, of *k* and *n*. Then draw KP parallel to OV, and O*p* cutting it in P, which will be the Original of the Point *p*. Laſtly drawing NP, you have the Original Figure ſought KLMNP.

DEMONSTRATION.

This Conſtruction is evident from *Problem 7.* it being the Reverſe of it.

N. B. In the ſame manner one may go back to the Original Figure by the Directors, as in *Fig.* 11.

PROBLEM X.

The ſame things being given, to find only the Length of the Original of a Projection given.

Let III (*Fig.* 10.) be the Projection given; the reſt of the Figure being to be underſtood as in the foregoing Problems. Continue III till it cuts the Vaniſhing Line in its Vaniſhing Point V. And draw VO. In the Vaniſhing Line take V3 equal to VO, and draw 3I and 3II cutting the Interſection in 1 and 2. Then will 1 2 be the Length ſought of the Original of III.

DEMON-

DEMONSTRATION.

Let W be the Interſection of I II. V 3 being equal to V O the Diſtance of the Vaniſhing Point V, and W 2 being parallel to V 3, the Point 3 may be conſider'd as the Point of Sight, and W 1.2 as the Original Line and 3.1 and 3.2, as Viſual Rays producing the Projection I II.

COROL. The Length 1.2 may be found by a Scale and Compaſſes, making 1.2 : V 3 (or to VO) : : I II × WV : IV × IIV.

*

PROBLEM XI.

Having given the Vaniſhing Line of a Plane, its Center and Diſtance, and the Projection of a Line in that Plane; to find the Projection of another Line in that Plane, making a given Angle with the former.

Let O (*Fig.* 10.) be the Point of Sight placed as in the foregoing Problems, GH being the Vaniſhing Line, and *ab* the given Projection; it being required to draw *ac*, ſo that the Original of the Angle *bac* may be equal to a given Angle. Continue *ab* to its Vaniſhing Point G. Draw GO, and OI making GOI equal to the given Angle, and cutting the Vaniſhing Line in I. Then draw *Iac*, which will be the Line ſought.

DEMON-

DEMONSTRATION.

The Figure being underſtood as in the foregoing Problems, let AB be the Original of *ab*, and conſequently be parallel to OG (by *Def.* 15. and *Theor.* 3.) For the ſame reaſon AC parallel to OI is the Original of *ac*, I being its Vaniſhing Point (by *Theor.* 3) But AB and AC being parallel to OG and OI, the Angle BAC is equal to GOI, which is equal to the given Angle by the Conſtruction. Wherefore *bac* repreſenting the Angle BAC, repreſents that given Angle. Which was to be done.

N.B. If it had been required to make *abc* to repreſent the Angle ABC, the Angle GOH muſt have been made equal to the Complement of the Angle ABC to two Right Angles.

PROBLEM XII.

Having given the Vaniſhing Line of a Plane, its Center and Diſtance, and the Projection of one Side of a Triangle of a given Species in that Plane; to find the Projection of the whole Triangle.

The Projections of the Sides wanting are to be found by the foregoing Problem, the Angles of the Triangle being given. Thus having given the Projection *ab* (*Fig.* 10.) of the Side AB of the Triangle ABC, the Vaniſhing Point I of the Side *ac* is found by making the Angle IOG equal to the Angle CAB, and the Vaniſhing Point H of of the Side *bc* is found by making the Angle HOG equal to the Complement of the Angle CBA to two Right Angles.

N. B. If the Vaniſhing Point of the Line given *ab* is out of reach, you may proceed thus. Taking any Line DR (parallel to the Vaniſhing Line HG by *Theor.* 6.) for the Interſection, by means of two Lines H*b*E, I*a*F, drawn at pleaſure thro' *b* and *a*, find the Originals A and B of the Points *a* and *b* (by *Prob.* 9.) and draw AB. Then on the Side AB compleat the Original Triangle, and find the Projections of the Sides wanting by *Prob.* 7.

PROBLEM XIII.

Having given the Vaniſhing Line of a Plane, its Center and Diſtance, and the Projection of one Side of any Figure in that Plane, to find the Projection of the Whole Figure.

Reſolve the whole Figure given into Triangles, by means of Diagonals, and find the Projections of thoſe Triangles one after another (by *Prob.* 12.) beginning with thoſe that have the Line given for one of their Sides.

The ſame thing may be done ſeveral Ways by the Application of the foregoing Problems, as is moſt convenient in every particular Caſe. This will be beſt underſtood by a few Examples.

Example I.

Fig. 14. In this Example IK is the Vaniſhing Line, S its Center, and SO its Diſtance, and AB parallel

parallel to IK is the given Projection of one side of a regular Hexagon. Having drawn OG parallel to IK (by *Def.* 15.) the Original of AB being parallel to the Picture (by *Cor.* 2. *Theor.* 5.) the Vanishing Points H,I,K of the Sides and Diagonals BC, FE, AD; AF, BE, CD; AC, are found by *Prob.* 11. making the Angles HOG 60 degr. IOG 120 degr. KOG 30 degr. Then drawing AK and BH, is got the Point C; drawing AH and CI is got D; drawing DE parallel to IK, and AS, is got E (for S is the Vanishing Point of AE, the Original of the Angle EAB being a Right Angle, as is GOS.) Lastly drawing EH and AI is got F, which compleats the Figure sought.

Example II. *

Fig. 15. In this Figure the Projection *mrptqs* of the Figure MRPTQS (which is the Ichnography of a Regular Icosaedron reposing on one of its Faces,) is found, having given the Projection *ab* of the Side AB, VX being the Vanishing Line, and O the Point of Sight reduced to the Picture, as in the foregoing *Problems.* The Original Ichnography is described by making two concentric and parallel regular Hexagons, AFBICH, and RMSQTP, whose homologous Sides are in the Proportion of the Parts of a Line cut in extream and mean proportion, (See *Def.* 3. *Lib.* 6. *Elem.*) and drawing the Lines, as is obvious enough in the Figure. *

Having continued *ab* to its Vanishing Point V, the Vanishing Points W and X of the other two Sides of the Triangle *abc* are found by

G *Prob.*

Prob. 11. Then drawing at pleasure s p parallel to V X, and drawing W *a* and W *b* cutting it in * a and b, and dividing a b in k, d, e, l, in the same proportion as A B is divided in K, D, E, L, and * draw kW, dW, eW, lW, are got the Projections *k*, *d*, *e*, *l* of the Points K, D, E, L, (by *Prob.* 3.) Then drawing *d* X, *e* W, and *e* X, are got the Points *f*, and *g*; and drawing *g* V, are got the Points *h* and *i*. Drawing *k* X and *l* W, is got the Point *m*, and *n* (which is the Projection of *N*) by the Intersection of *k* X with *d* W already drawn. Then making d o and o p each equal to m d, and drawing o W, p W, are got the Points *o* and *p* (by *Prob.* 3.) Then drawing *p* V is got the Point *q*, by its Intersection with *l* W already drawn. Drawing *o* W and *n* V is got *r*: And making m s equal to m d, and drawing s W is got *s*. Lastly, drawing *s* X cutting *r o* W already drawn, is got *t*. The rest is done by joining the Points found, as is evident enough in the Scheme.

Example III.

Fig. 16. In this Example is found the Projection of the Ichnography of a regular Dodecaedron, having the Projection of one Side given, by returning to the Original Figure, by *Prob.* 9. and then proceeding by *Prob.* 7. I shall leave the Reader to exercise himself in considering this * Scheme, and only observe, that the Original Ichnography is made by describing two concentric and parallel Decagons, whose homologous Sides are as the Segments of a Line cut in extream and mean Proportion.

Example

Example IV.

Fig. 17. In this Figure DC being the Vanifhing Line, and O the Point of Sight reduced to the Picture, as in the foregoing Problems, the Projection ANBMLP, of a regular Octaedron, * having given the Projection of one of the Sides AB, is found as follows. Having continued AB to its Vanifhing Point C, the Vanifhing Point G of the Sides AK and NM is found by *Prob*. 11. making the Angle COG equal to 60 degr. Then (by *Prob*. 10.) taking any Line *bl* (parallel to CD) for the Interfection, and making CD equal to CO, and GH equal to GO, and drawing DA and DB cutting *bl* in *a* and *b*, *ab* is the Length of the Original of AB, and drawing HA cutting *bl* in a, and making a*l* equal to *ab*, and drawing *l*H cutting AG in L, is got the Projection AL (the Original of it being equal to a*l*, and confequently to the Original of AB, a*l* and *ab* being equal. Then dividing *ab* and a*l* each into three equal Parts by the Points *e*, *f*, *i*, *k*, and drawing *e*D, *f*D, *i*H, *k*H, are got the Points E, F, I, K (by *Prob*. 3.) Then drawing FG, KH, EI, are got the Points M, N, P, which compleat the Figure.

PROBLEM XIV. *

Having given the Center and Diftance of the Picture, and the Vanifhing Line of a Plane; to find the Vanifhing Point of Lines perpendicular to that Plane.

 Let

Let AB (*Fig.* 18.) be the Vaniſhing Line given, and C the Center of the Picture. Draw CA perpendicular to AB, and CO parallel to it, and equal to the Diſtance of the Picture. Draw AO and OD perpendicular to it, cutting CA in D; which will be the Vaniſhing Point ſought.

DEMONSTRATION.

Imagine the Triangle AOD to be turned up on the Plane of the Scheme, ſo that CO may be perpendicular to it, O being brought into the Point of Sight. This being done, the Plane paſſing through the Point O and the Line AB will be the Parallel of the Original Plane, and the Line OD will be perpendicular to it; and conſequently will be the Parallel of Lines perpendicular to that Original Plane. Wherefore D is the Vaniſhing Point of thoſe Perpendiculars (by *Def.* 17.)

N.B. 1. When the Vaniſhing Line AB paſſes through the Center of the Picture, that is, when the Original Plane is perpendicular to the Picture, the Point D will be infinitely diſtant, the Line OD being parallel to AD, and the Projections of the Lines perpendicular to the Plane propoſed will all of them be perpendicular to AB, they being to meet the Line AC, which is perpendicular to it at an infinite diſtance; and conſequently they will be parallel to one another. Which they ought to be upon another account, their Originals being all parallel to the Picture.

N.B. 2. But when the Original Plane is parallel to the Picture, the diſtance CA will be infinite, and conſequently OA will be parallel to

CA,

CA, and OD will coincide with OC, making the Point D to fall into the Center of the Picture C, agreeably to *Cor.* 2. *Theor.* 3.

N.B. 3. CD is a third proportional to AC and CO; as alſo is AD to AC and AO.

N.B. 4. OD is the Diſtance of the Vaniſhing Point D.

PROBLEM XV. *

Having given the Center and Diſtance of the Picture, to find the Vaniſhing Line, its Center and Diſtance, of Planes that are perpendicular to thoſe Lines that have a certain given Vaniſhing Point.

Let C (*Fig.* 18.) be the Center of the Picture, and D the Vaniſhing Point given. Draw DC, and CO perpendicular to it, and equal to the Diſtance of the Picture. Then draw DO, and OA perpendicular to it, cutting DC in A. Perpendicular to DC draw AB, which will be the Vaniſhing Line ſought, A being its Center (by *Theor.* 1.) and OA its Diſtance.

This Conſtruction follows neceſſarily from the Conſtruction of the foregoing Problem, and the Obſervations at the End of that may be applied here.

PRO-

* PROBLEM XVI.

Having given the Center and Diſtance of the Picture, through a given Point to draw the Vaniſhing Line of a Plane that is perpendicular to another Plane whoſe Vaniſhing Line is given, and to find the Center and Diſtance of that Vaniſhing Line.

Let AB (*Fig.* 18.) be the Vaniſhing Line given, and C the Center of the Picture, and let E be the Point given. Find the Vaniſhing Point D of Lines perpendicular to the Original Planes of AB, (by *Prob.* 14.) Draw DE, which will be the Vaniſhing Line ſought. Draw CF cutting DE at Right Angles in F, and F will be the Center of the Vaniſhing Line DE (by *Theor.* 1.) Make a Right-angled Triangle, whoſe Baſe is CF, and its Perpendicular is equal to the Diſtance of the Picture, and its Hypothenuſe will be the Diſtance of the Vaniſhing Line DE (by *Cor. Th.* 1.)

DEMONSTRATION.

Becauſe the Plane, whoſe Vaniſhing Line is ſought, is perpendicular to the other Plane, its Vaniſhing Line muſt paſs thro' the Vaniſhing Point D of Lines perpendicular to that other Plane, becauſe ſome of thoſe Lines are in the Plane ſought. Therefore DE is the Vaniſhing Line ſought. The reſt needs no Demonſtration.

COROL.

COROL. 1. If FC be continued till it cuts the Vaniſhing Line given in B, B will be the Vaniſhing Point of Lines perpendicular to the Original Plane of the Vaniſhing Line DE. For that Vaniſhing Point is in the Line FC by the Conſtruction of *Problem* 14. and it is in the Vaniſhing Line given by the Demonſtration of the preſent Problem.

COROL. 2. And therefore if the Vaniſhing Lines AB and DE meet in G, the Points B, D, and G will be the Vaniſhing Points of the three Legs of the ſolid Angle of a Cube, which are perpendicular to one another. And drawing DB; BG, GD, and DB will be the Vaniſhing Lines of the three Planes that contain that ſolid Angle.

COROL. 3. The Diſtance of the Vaniſhing Line DG is equal to the Line FP, the Point P being the Interſection of the Line FC with a Circle deſcribed on the Diameter DG.

PROBLEM XVII. *

Having given the Center and Diſtance of the Picture, and the Vaniſhing Point of the common Interſection of two Planes that are inclined to one another in a given Angle, and the Vaniſhing Line of one of them; to find the Vaniſhing Line of the other of them.

Let C (*Fig.* 18.) be the Center of the Picture, BG the given Vaniſhing Line of one of the Planes, and B the Vaniſhing Point of their common

mon Interſection, and H the Angle of their Inclination to one another. Find the Vaniſhing Line GD, of Planes perpendicular to the Lines whoſe Vaniſhing Point is B (by *Prob.* 15.) Let that Vaniſhing Line cut the Vaniſhing Line given in G. In GD find the Vaniſhing Point E of Lines making the given Angle H with the Lines whoſe Vaniſhing Point is G (by *Prob.* 11.) that is, in BCF perpendicular GFD take FP equal to the Diſtance of the Vaniſhing Line GD (found by *Prob.* 15.) and draw PG, and PE making the Angle EPG equal to H. Draw BE, which will be the Vaniſhing Line ſought.

DEMONSTRATION.

Imagine the Triangle GPE to be turned up on the Line GE, ſo that the Point P may be in the Point of Sight perpendicular over the Center of the Picture C. In that caſe the Planes BPG, * GPD, DPB will be the Parallels of three Original Planes, whoſe Vaniſhing Lines are BG, GD, * DB ; that whoſe Vaniſhing Line is DG being perpendicular to the other two (by the Conſtruction, becauſe it is perpendicular to their common Interſection, whoſe Vaniſhing Point is B.) Therefore the Original Planes, whoſe Va- * niſhing Lines are BG and BD, are inclined to one another in the Angle EPG, that is, in the Angle H, (for the Inclination of two Planes is always meaſured in a Plane perpendicular to their common Interſection.) Therefore BG being the Vaniſhing Line given, BE is the Vaniſhing Line ſought.

N.B.

N.B. The Center of the Vaniſhing Line BE is found by drawing a Line perpendicular to it from C (by *Theor.* 1.) and then its Diſtance is found, as was found the Diſtance PF in *Prob.* 16.

PROBLEM XVIII.

Having given the Center and Diſtance of the Picture, and the Vaniſhing Line of one Face of any ſolid Figure propoſed, and the Projection of one Line in that Face; to find the Projection of the whole Figure.

By means of the Projection given find the Projection of the Ichnography of the Figure propoſed on the Plane of that Face, whoſe Vaniſhing Line is given (by *Prob.* 13.) Then (by *Prob.* 16.) find the Vaniſhing Line of the Plane of the Orthography, and deſcribe the Projection of the Orthography by the help of the Lines already given in the Ichnography (by *Prob.* 13.) Laſtly, by the interſections of the Projections of Perpendiculars to the Ichnography and Orthography, will be found the ſeveral Points of the Projection required.

Otherwiſe. Having found the Projection of the Face whoſe Vaniſhing Line is given, by means of the Projection of the Line given, find the Vaniſhing Lines of the adjacent Faces (by *Prob.*17.) and deſcribe their Projections by the help of the Lines given in the Projection of the firſt Face; and ſo on, till the whole Projection ſought is compleated.

This is a general Defcription of the Method to be ufed in putting any Figures propofed into *Perfpective*; but in the Practice of particular Cafes feveral Expedients may be ufed, in the various Application of the foregoing Problems. But all this will be beft underftood by Examples.

* *Example* V.

Fig. 19. In this Figure is found the Projection of a regular Dodecaedron, having given the Projection AB of one Side parallel to the Picture, by means of the Ichnography and Orthography; FG (parallel to AB) being the given Vanifhing Line of the Face ABCDE, F its Center, H the Center of the Picture, and HO its Diftance. To avoid Confufion of Lines, the Ichnography and Orthography are removed from the Space taken up by the Projection fought in the following manner. For the Ichnography; drawing at pleafure *ab* parallel to AB, and at a fufficient diftance from it, and then drawing IA and IB cutting *ab* in *a* and *b*, the Line AB is transfer'd to *ab*; I being the Vanifhing Point of Lines perpendicular to the Face ABCDE, whofe Vanifhing Line is FG (found by *Prob.* 14.) This being done, the whole Ichnography is defcribed on the Line *ab*, by *Prob.* 13. as it is defcribed in *Ex.* 3.

For the Orthography; IHF paffing thro' the Center of the Picture, is taken for its Vanifhing Line, as being moft convenient; the Orthography in this cafe being the moft fimple, and the Projections of Lines perpendicular to it being all of them perpendicular to FI (by *Note* 1. *Prob.* 14.) Having drawn GA (from the Vanifhing

4 Point

Point G of the Line *a e* in the Ichnography) and *e*I cutting it in E, is got the Projection A E. Then drawing A a and E e both parallel to G F (and consequently representing perpendiculars to the Orthography, as is already said) and a e at pleasure passing thro' F, and cutting them in a and e, is got a e; by the means of which the whole Projection of the Orthography is described (by *Prob.* 13.)

Having thus got the Projections of the Ichnography and Orthography, any Point K of the Projection sought is got by drawing k K parallel to F G and *k* I, from the corresponding Points k and *k*, meeting each other in K.

For understanding the Projection of a Cube, which appears in this Figure, after what has been already said, it is sufficient to inform the Reader, that two of its Vanishing Lines are F G, and F I, and the third is a Line passing thro' I parallel to F G.

As to the Shadows, which are supposed to be cast by the Sun on the Plane of the Face ABCDE of the Dodecaedron; the Shadow *u v* of any Line V *v* is found as follows. S is the given Vanishing Point of all the Rays of Light, which being supposed to come from the Sun, are to be consider'd as parallel. Therefore IS passing thro' the Vanishing Point I of the Line V *v* and the Vanishing Point S of the Rays, is the Vanishing Line of the Plane made by all the Rays passing thro' the Line V *v*, and projecting the Shadow *v u*. And IS cutting the Vanishing Line F G of the Plane the Shadow is cast on in *s*, *s* is the Vanishing Point of the Shadow *v u*, which is the

 Inter-

Interſection of the Plane of the Shade (whoſe Vaniſhing Line is I*s*) and the Plane the Shadow is caſt on (*s* being the Vaniſhing Point of that Interſection by *Cor.* 2. *Th.* 7.) Having therefore drawn *vs*, and VS cutting it in *u*, *vu* is the Shadow of the Line *v*V.

* *N.B.* In the Original of the Orthography in the preſent Figure, the Points e, m, s, p are the Angles of a Square. The Lines o n, a l, q r being parallel to e m, and o q, n r being parallel to e p, and equal to a l. t k, o n, q r, are equal, and o n, e m, a l, are in the continued Geometrical Proportion, of the leſſer Segment to the greater, of a Line cut in extream and mean Proportion.

Several Lines mention'd in this Scheme are not actually drawn, to avoid Confuſion.

Example VI.

Fig. 20. In this Scheme the Vaniſhing Line of the Ground, which the Piece of Building, &c. ſtand upon, is AICKB paſſing through the Center of the Picture C, the Diſtance of the Picture being equal to CO. BG and AH are the Vaniſhing Lines of the upright Planes a b d e, DE N, &c. and *abc*, DF *nm*, &c. G and H being the Vaniſhing Points of the Lines *be*, and *mn*, which touch the upper Corners of the two Flights of Steps, and conſequently AG and BH are the Vaniſhing Lines of the Planes that touch the upper or the under Edges of the Steps.

The given Side *hr* of the Baſe of the regular Tetraedron is parallel to the Vaniſhing Line AB. *u*, which is the Projection of the Center of the
* Baſe *hrp*, and the Seat of the Vertex *o*, is found by

by drawing *pq* parallel to AB, and I*r* cutting it in *q*, and then drawing *Cp* and *hq* meeting in *u*. CL perpendicular to AB is the Vaniſhing Line of a Plane perpendicular to the Baſe *krp* ſtanding on *up*, and paſſing thro' the Line *po*, whoſe Vaniſhing Point is L, found (by *Prob.* 11.) by making CQ equal to the Diſtance of the Picture, and the Angle LQC equal to the Original of the Angle *upo*. VX is the Vaniſhing Line of the Face *upr*, by the help of which that Face is deſcribed, having given *pr* (by *Prob.* 12.) as the Face *hrp* was deſcribed on *hr*. The regular Octaedron, and Icoſaedron, are deſcribed by their Ichnography and Orthography, which Method is ſufficiently explain'd in the foregoing Example. I will only inform the Reader, that in the Orthography ABCDEFGH of the Icoſaedron, the Lines CD, AF, GH are equal, as alſo are AC, FD, BE, and AF is to BE, as the leſſer Segment is to the greater of a Line cut in extream and mean Proportion. * * * * *

The Light is ſuppoſed to come from the Sun, and the Rays are parallel to the Picture and to the Line AM; ſo that the Shadow P of any Point D, is found by drawing NP through its Seat parallel to AB, and DP parallel to AM cutting it in L. For the Shadow of the Tetraedron, which is turned up on the End of the Steps; having in the ſame manner found *s*, (which would be the Shadow of *o* on the Ground, if the Steps were away) drawing *sh* and *sp*, are got the Shadows *ho* and *po*. Let *su* and *sp* cut the lower Edge of the Steps in *y* and *x*; then drawing *xt* perpendicular to AC and cutting *os* in *t*, is got the Shadow

Shadow *t* of the Vertex *o* on the End of the Steps, and drawing *xt* is got that part of the Shadow of *op*, which falls on that End. And in the same manner is got the Shadow of *ob*. All the Rays being parallel to AM, and A being the Vanishing Point of DF, AM is the Vanishing Line of the Plane made by the Rays which pass thro' the Line DF, and BM being the Vanishing Line of the Wall, the Shadow F*f* is cast on; M is the Vanishing Point of the common Intersection of those two Planes, that is, of the Shadow F*f* of the Line DF.

Example VII.

Fig. 21. In this Scheme ACB passing through the Center of the Picture, C is the Vanishing Line of the Ground; A is the Vanishing Point of the Edge EF which the Beam rests upon, and the other Edges parallel to it; D is the Vanishing Point of the Edge GH, &c. and DB perpendicular to AB, is the Vanishing Line of the Plane EGH; P and Q are the Vanishing Points of the Sides MK and KN, of the regular Tetraedron, and consequently PQ is the Vanishing Line of the Triangle KNM; L is the Light and A its Seat on the Ground. Having BE*g*, that is the intersection of the upright Plane EGH with the Ground, and consequently *g*, where HG meets it, is the intersection of the Edge HG with the Ground. BE cuts the Edge SR of the Wall SRh in R, and consequently R*h* perpendicular to AB is the intersection of the Wall with the Plane EGH, and consequently *h*, where R*h* and GH meet, is the intersection of the Line GH

GH with the Wall. DL*l* is the Projection of a Line parallel to the Original of GH (becauſe of the Vaniſhing Point D) and Bl is its Seat, wherefore *l* is its interſection with the Ground.

Now the Originals of L*l* and HG being parallel, they are in the ſame Plane, *viz.* in the Plane which makes the Shadow of HG; but *lg* repreſents the interſection of that Plane with the Ground; wherefore *lg* continued is part of the Shadow, and drawing LG cutting it in g, g is the Shadow of G, and gS is that part of the Shadow of HG which is on the Ground. Then drawing S*h* and LH cutting it in h, hS is the other part of that Shadow againſt the Wall.

Having drawn *gp* and DT both parallel to AB, D*g* and *gp* are the Projections of two Lines in a Plane, whoſe Vaniſhing Line is DT; and T is the Vaniſhing Point of the common interſection of that Plane with the Plane of the Triangle KMN (by *Cor.* 2. *Th.* 7.) PQT being its Vaniſhing Line. Therefore drawing Tp cutting gD in V, V is the interſection of the Line *g*GHD with the Plane of that Triangle KMN. Then *lg* cutting MK in *r*, drawing *rt*V, *rt* is that part of the Shadow of GH, which falls on the Triangle KMN.

Having thus explain'd the manner of finding the Shadow of GH, the reſt needs no Explanation.

Example VIII.

Fig. 22. In this Scheme C is the Center of the Picture, and CA the Vaniſhing Line of the Ground, and of the Surface of the Water which gives

gives the Reflexions; and S is the Vanifhing Point of the Rays of Light, which are fuppofed to come from the Sun.

The Shadow of the perpendicular Line BD is found thus: SA drawn perpendicular to CA gives the Vanifhing Point A of the Shadow on the Ground B*f*. Then in the Circumference of the Bafe of the Cylinder (which is parallel to the Picture, its Axis being perpendicular to it, and confequently having the Vanifhing Point C) taking any Point E, and finding its Seat on the Ground *e*, drawing CE, and C*e* cutting BA in *f*, and then drawing *f*P perpendicular to CA and cutting CE in P, is got one Point P of the Shadow on the Surface of the Cylinder. And in the fame manner are got all the other Points of that Shadow. To prove this, the Reader need only confider, that the Original of *e*EP*fe* is an upright Plane cutting the Cylinder in EP, and *f*P is the Shadow of BD on that Plane. Wherefore P is the Point where that Shadow falls on the Surface of the Cylinder. Any Point Q of the Shadow of the Circumference of the inward Cylinder on its Surface, is found thus. Having drawn CS, parallel to it is drawn at pleafure GH cutting that Circumference in G and H. Then drawing GS and CH meeting in Q, Q is the Point fought. For C being the Vanifhing Point of the Axis of the Cylinder, as well as of CH, HQ is in the Surface of the Cylinder, and GH being parallel to CS, is the Projection of a Line parallel to the Picture, in a Plane whofe Vanifhing Line is CS (and its Original being parallel to the Picture, is in the Bafe of the Cylinder, which is parallel

parallel to the Picture.) Therefore the Originals of HG, HC, and GS being in the ſame Plane, Q is the Projection of the Point where the Ray of Light, whoſe Projection is GS, cuts the Surface of the Cylinder; that is, it is the Projection of the Shadow of the Original of the Point G in the Circumference of the Baſe, on the inward Surface of the Cylinder.

b being the Seat of the Point D on the Surface of the Water, the Reflexion *d* of the Point D is found by continuing the Perpendicular D*b* till *bd* is equal to *b*D. This is evident, becauſe the known Law of Reflexions, is, that the Reflexions of all Objects appear to be as much on one Side of the reflecting Plane, as the real Objects are on the other Side of it. In AS, making A*s* equal to AS, any Point *q* in the Shadow on the Surface of the inward Cylinder in the Reflexion, is found in the ſame manner as Q in the real Figure, uſing the Point *s* inſtead of S.

The Shadow of the Cylinder on the Surface of the Cone, is found by ſuch another Expedient, as the Shadow of the Line BD on the Surface of the Cylinder.

Example IX. *

Fig. 23. In this Scheme every thing elſe being eaſily to be underſtood by what has been already explain'd, I ſhall only ſhew the manner how the Reflexion is found in the Looking Glaſs of the Picture on the Eazle.

A is the Center of the Picture, and AB the Vaniſhing Line of the Ground; the Diſtance of the Picture being equal to AB. AC is the Va-

I niſhing

nishing Line of the Picture on the Eazle, and CD the Vanishing Line of the Looking Glass.

Through *a*, where the Edge *b a* of the Leg of the Table cuts the Surface of it, drawing *a e*, and through *b* drawing *b d*, both parallel to AB, *b d* cutting the Intersection *c d* of the Surface of the Picture on the Eazle with the Ground in *d*, and then drawing *d e* parallel to AC, and cutting *a e* in *e*, and then drawing A *e*, is got the Projection A *e* of the common Intersection of the Surface of the Table, and of the Picture on the Eazle. For *a e* being parallel to AB, is the Projection of a Line in the Surface of the Table parallel to the Picture, and for the same reason *b d* is the Projection of a Line on the Ground, and *d e* is the Projection of a Line in the Plane of the Picture on the Eazle, both of them parallel to the Picture; *a b* is also the Projection of a Line parallel to the Picture. Therefore *a b d e* is the Projection of a Trapezium parallel to the Picture, whose Angle *e* is in the common intersection of the Surface on the Table, and of the Picture on the Eazle. But A being the common intersection of the Vanishing Lines of those two Planes, is the Vanishing Point of their common intersection, and therefore *e* A is the Projection of that intersection (by *Cor.* 2. *Th.* 7.) For the same reason *o* being the Projection of the Point where the Surface of the Glass touches the Table, and E being the common intersection of the Vanishing Lines AB and CD, *o* E is the Projection of the common intersection of the Surface of the Table and the Surface of the Glass. Therefore *f* where *o* E and *e* A meet, is the Projection of the Point where

where the three Planes meet, of the Surface of the Table, the Glaſs, and the Picture on the Eazle. Therefore drawing *f* C, it is the Projection of the common interſection of the Picture on the Eazle and the Looking Glaſs.

Having found the Vaniſhing Point P of Lines perpendicular to the Plane of the Looking Glaſs, whoſe Vaniſhing Line is CD (by *Prob.* 14.) drawing PA thro' the Vaniſhing Point A of the Line GH, and cutting CD in D, D is the Vaniſhing Point of the Seat of GH on the Plane of the Glaſs. Therefore GH cutting C*f* in *i*, D*i* is the Projection of that Seat. Then drawing GP cutting D*i* in *k*, *k* is the Seat of the Point G on the Glaſs. Wherefore in GP making *kg* to repreſent a Line equal to that repreſented by G*k* (by *Prob.* 3.) *g* is the Projection of the Reflection of G, and *gi* is the Reflexion of G*i*, and drawing PH cutting *gi* in *h*, *gh* is the Reflexion of GH. And in the ſame manner may be found any other Lines in the Reflexion.

The Reflexion of the Picture on the Eazle may alſo be deſcribed by its Vaniſhing Line, in the ſame manner as the Projection of the Picture it ſelf was deſcribed; for in PAD making *a*D to repreſent a Line equal to that repreſented by AD, *a* is the Vaniſhing Point of the reflected Line *gh*, and C*a* is the Vaniſhing Line of the reflected Picture on the Eazle.

LINEAR PERSPECTIVE.

PART II.

Of the Manner of finding the Original Figures from their Projections given, and of the Situation that is neceſſary to be obſerved for viewing particular Projections.

PROBLEM XVIII.

Having given the Projection of a Line divided, and its Vaniſhing Point; to find the Proportion of the Parts of the Original.

LET AB (*Fig.* 7.) be the given Projection, divided in C, and V its Vaniſhing Point. Draw at pleaſure VO, and *ab* parallel to it, and from any Point O in the Line OV draw OA, OB, OC, cutting *ab* in *a*, *b*, and *c*. Then will

will the Original of AC be to the Original of CB, as *ac* is to *cb*.

COROL. *ac* : *cb* :: AC×BV : BC×AV.

PROBLEM XIX.

Having given the Projection of a Line divided into two Parts, and the Proportion of the Original; to find its Vanishing Point.

Let AB (*Fig.* 17.) be the Projection given divided in C. Through C draw at pleasure aCb, and in it make aC to Cb, as the Original of AC is to the Original of CB, and draw aA and bB meeting in O. Parallel to ab draw OV cutting AB in V, which will be the Vanishing Point sought. *

COROL. BV : BA :: Ca×CB : Cb×AC —Ca×CB.

These two last Problems, with their Corollaries, follow easily from *Prob.* 3. and its *Cor.*

PROBLEM XX.

Having given the Projection of a Triangle, with its Vanishing Line, its Center and Distance; to find the Species of the Original Triangle.

Let *abc* (*Fig.* 10.) be the Projection given, HG its Vanishing Line, and S its Center, and SO perpendicular to HG, and equal to its Distance. Having continued the Sides of the Projection given,

given, till they cut the Vaniſhing Line in their Vaniſhing Points G, H, I, draw GO, GH, and GI, and the Originals of the Angles *bac*, *ab*H, *acb*, will be equal to GOI, IOH, GOH, reſpectively (by *Prob.* 11.) Whence the Species of the Original Triangle is given.

PROBLEM XXI.

Having given the Projection of a Triangle of a given Species, and its Vaniſhing Line; to find the Center and Diſtance of that Vaniſhing Line.

Let ABC (*Fig.* 12.) be the given Projection, and FD its Vaniſhing Line. Continue the Sides of the Projection till they cut the Vaniſhing Line in their Vaniſhing Points D, E, F. Biſect DE and EF in G and H, and draw GI and HK perpendicular to FD, making GI to GE as Radius is to the Tangent of the Angle repreſented by BAC, and KH to EH as Radius is to the Tangent of the Angle repreſented by BCA; ſo that EIG and FKH may be equal to thoſe Angles. With the Centers I and K and the Radius's IE and KE, deſcribe two Circles cutting each other in O, and draw OS cutting FD at Right Angles in S. Then will S be the Center, and SO the Diſtance ſought.

DEMONSTRATION.

Suppoſing S to be the Center and SO the Diſtance of the Vaniſhing Line FD, the Originals of the Angles BAC and BCA will be equal to DOE, and EOF (by *Prob.* 11.) But by the nature

nature of the Circle DOE and FOE are equal to GIE and HKE, which by the Conſtruction are equal to the Angles that ought to be repreſented by BAC and BCA. Therefore S is the Center and SO the Diſtance ſought.

PROBLEM XXII. *

Having given the Projection of a Trapezium of a given Species; to find its Vaniſhing Line, Center and Diſtance.

Let *abcd* (*Fig.* 12.) be the Projection given. Draw the Diagonals *ac*, *bd*, meeting in *e*, and by the Proportions of the Originals of *ae*, *ec*, and *be*, *ed*, find the Vaniſhing Points E and F, of the Lines *ac* and *bd* (by *Prob.* 19.) Draw FE, which will be the Vaniſhing Line ſought. Then by the given Species of the Original of the Triangle *abe*, find the Center S and Diſtance SO (by *Prob.* 21.)

PROBLEM XXIII *

Having given the Projection of a Right-angled Parallelopiped; to find the Center and Diſtance of the Picture, and the Species of the Original Figure.

Let ABCDEFG (*Fig.* 24.) be the Projection given. Continue the Projections of the parallel Sides, till they meet in their Vaniſhing Points H, I, K, and draw HI, HK, IK, which will be the Vaniſhing Lines of the ſeveral Faces of the Figure ſought,

ſought, containing a ſolid Right Angle. Draw KL perpendicular to HI, and HM perpendicular to KI meeting in S; which will be the Center of the Picture (by *Cor.* 2. *Prob.* 16.) Then on the Diameter LK deſcribe a Circle, and draw SO perpendicular to LK cutting it in O, and OS will be the Diſtance of the Picture (by *Prob.* 14. LOK being a Right Angle upon account of the Circle.) Laſtly, find the Diſtances of the Vaniſhing Lines KI and IH (by *Cor.* 3. *Prob.* 16. M and L being their Centers, by *Th.* 1.) and then find the Species of the Originals of the Faces DAFE, and DABC (by *Prob.* 20.)

COROL. When the Vaniſhing Line of one of the Faces (ſuppoſe IH) paſſes through the Center of the Picture, the Vaniſhing Point K of the Sides perpendicular to it, will be at an infinite diſtance: by which means the Situation of LK will be indetermined. So that the Species of that Face ABCD may be taken at pleaſure, and then the Center and Diſtance of the Picture may be found by *Prob.* 20. And in this caſe, if it were only required that the Projection propoſed ſhould repreſent a right angled Parallelopiped in general, the Place of the Point of Sight might be any where in the Circumference of a Circle deſcribed on the Diameter HI, and in a Plane perpendicular to the Picture. This I leave as a hint that may be uſeful to the Painters of Scenes in Theaters.

F I N I S.

APPENDIX.

NUMB. I. *

The Deſcription of a Method, by which the Repreſentations of Figures may be drawn on any Surface, be it never ſo irregular.

FROM what has been ſaid in this Book to explain the Principles of Painting, eſpecially at the End of the Definitions, it is evident that the Senſe of the 2[d] Theorem may be extended to any Surface that Figures are painted on, be it Concave or Convex, or never ſo irregular. So that be the Surface of the Picture ABC (*Fig* 3.) of any Form whatſoever, the Projection *fg* of the Original Line FG, will ſtill be the interſection of the Picture with the Plane of the Triangle FGO. But the Parallel OV is in that Plane (*Th.* 3.) Hence it follows, that if the

K Plane

Flame of a Lamp be so placed, as to cast the Shadow of OV on any Point B of the Line FG, it will cover the whole Line FG, and at the same time cover the Line V*fg*B on the Picture, which is the Projection of FG; all the Rays coming from the Lamp, and passing by the Line OV, in that case making a Plane, which coincides with the Plane of the Triangle OFG. The same thing will happen, if a Person should place his Eye so as to make any Point of the Line FG to seem to be cover'd by the Line OV; in that case OV will seem to cover all the Projection V*fg*. So that if a Lamp be so placed, as to make the Shadow of OV to pass thro' any one Point of the Projection V*g*, it will coincide with the whole; and if a Person places himself so as to make OV to seem to cover any one Point of the same Projection, it will seem to cover the whole. Hence I imagine, that the following Method may be of use for drawing the Projections of any Figures on any Surface, suppose on the Walls and Cupola's of Churches; the Walls and Cielings of great Rooms, the Scenes of Theaters, &c.

Chuse some principal Line in the Design to be drawn, and having by some proper Method found the Projections of its extream Points, through the Point of Sight pass a Thread parallel to the Original Line, and cast the Shadow of it on those two Points already found, and that Shadow mark'd with a Crion will be the Projection of that principal Line; or, if in any particular case it happens to be more convenient, place your Eye so as to make that parallel Thread to seem to cover those two Points already mark'd, and instruct

inſtruct an Aſſiſtant to mark out the Projection wanted. Suppoſe, for example, *ph* (*Fig.* 20.) to be that principal Projection. Then to find the Projection (for example *o*) of any other Point in the Figure to be deſcribed, imagine that Point to be the Vertex of a Triangle, whoſe Baſe is the Original of the Projection already found, and place the Thread (paſſing ſtill through the Point of Sight) in a parallel Situation to the Original of one of the Legs of that Triangle, and caſt its Shadow on the proper Extremity of the Projection given, and mark it as before, and you will have the indefinite Projection of that Leg, (ſuppoſe it to be *ho.*) Do the ſame by the other Leg, and by the interſection of thoſe Projections (ſuppoſe of *ho* and *po*) you will have the Projection of the Point ſought. And by this Method may be found the Projections of any Figures whatſoever. I ſhall not enlarge upon this Method, not having had an opportunity of putting it in practice; for which reaſon I only propoſe it as a Hint, which I leave to be further conſider'd of by the Curious.

*

NUMB. II.

A New THEORY for mixing of COLOURS, *taken from Sir* Isaac Newton's Opticks.

THO' my Design was only to treat of *Linear Perspective*, and not to discourse of all the Parts of *Painting*, yet as this Book will be most useful to those who practise that Art, I thought it would not be improper, nor unentertaining to the Readers, if I took this Occasion to publish some Thoughts I have had concerning the Mixture of Colours: which I have fallen into upon considering Sir *Isaac Newton*'s Theory of Light and Colours, in his most excellent Treatise of *Optics*.

In Colours these two things are to be consider'd, the Hue, (which is properly what may be called the Colour,) and the Strength of Light and Shadow. For as different Colours, suppose Red and Green, may have the same Strength of Light; so two things, that are one of them much darker than the other, may still have the same Hue, as a light Blue and a dark Blue.

With

With respect to the Hue, these two things are to be consider'd, 1st. The Species of Colour, and 2dly. The Perfection and Imperfection of Colour under the same Species. Colours differ in Species, as Blue and Red, and Colours of the same Species differ in degree of Perfection; as the Red of a Field Popy is much more perfect than the Red of a Brick. This Quality of Perfection and Imperfection in the Colours, by the Painters is express'd by the Terms Bright, or Clean, or Simple, and Broken; which is taken from their Method of making the imperfect Colours, by the Mixture of other Colours, which is called breaking the Colours. With respect to this Quality of Colours, Sir *Isaac Newton*, in the Book already mention'd, shews, that every Ray of Light has its proper Colour, which it always carries with it, and never loses, in whatever manner it happens to be reflected or refracted. These natural Colours of the Rays are the Bright Simple Colours, and the natural Order of them, as they appear when they are separated by the Refraction of a Prism, is, Red, Orange, Yellow, Green, Blue, Indico, Violet. All the less perfect or broken Colours, are made by the Composition and Mixture of these simple Colours, as Yellow Rays mix'd with Blue Rays, make a Green, but not so perfect as the simple natural Rays that are Green; and Red and Yellow Rays make an Orange Colour, but not so perfect as the natural Orange-colour'd Rays. And by a just Proportion of all the Natural Rays together, is produced Whiteness, which is indifferent to all the simple

ſimple Colours, and can't be ſaid to incline more to one Colour than to another. By White I mean any Colour between the lighteſt White and the darkeſt Black ; for as we are not now conſidering the Degrees of Light and Shade, all the Colours from Black to White are to be conſider'd as of the ſame Hue.

According to this Obſervation of the Nature of Whiteneſs, it appears that the broken Colours are a Medium between the ſimple Colours and White, and the more broken a Colour is, the nearer it is to White, and the further it is from White, the more ſimple it is.

Having thus explain'd the Nature of the Colours, and the Effect of their Mixture, in order to find exactly what Colour will be produced by the Mixture of any Colours given, Sir *Iſaac* diſpoſes the Colours in the following manner. Let there be a Circle made ADFA, and let the Circumference be divided into ſeven Parts AB, BC, CD, DE, EF, FG, GA, in the ſame proportion to one another as the Fractions $\frac{1}{9}, \frac{1}{16}, \frac{1}{10}, \frac{1}{9}, \frac{1}{10}, \frac{1}{16}, \frac{1}{9}$; which are the Proportions of the Muſical Notes *Sol, la, fa, ſol, la, mi, fa, ſol.* Between A and B place all the Kinds of Red, from B to C place all the Kinds of Orange, from C to D place all the Kinds of Yellow, from D to E place all the Kinds of Green, from E to F place all the Kinds of Blue, from F to G place all the Kinds of Indico, and from G to A place all the Kinds of Violet. Having thus diſpoſed the ſimple Colours, the Center of the Circle O will be the Place of White. And between the Center and the Circumference are

are the Places of all the broken compounded Colours, thoſe neareſt the Center being the moſt compounded, and thoſe fartheſt from it being the leaſt compounded. As in the Line O 1, all the Colours at 1, 2, 3, 4, are of the ſame Species, that is, Green inclining towards Blue, but the Colour at 1 is the ſimple natural Colour; that at 2 is ſomething compounded, or broken; that at 3 is more broken; and that at 4 is ſtill more broken.

The Colours being thus diſpoſed, to know what Colour reſults from the Mixture of any Colours given, find the Center of Gravity of the Places of the Colours given, and that will ſhew the Character of the Compound. For example, ſuppoſe I would know what Colour would reſult from the Mixture of two Parts of the ſimple Yellow at P, with three Parts of the ſimple Blue at Q: I find the Center of Gravity 3 of the Points P and Q; that is, I draw PQ, and having divided it into five Parts (which is the Sum of three and two) I take the Point 3 three Parts from P (becauſe there are three Parts of Blue) and two Parts from Q, (becauſe there are two Parts of the Colour at P.) Then drawing O 3 cutting the Circumference in 1, by the Place of the Point 1, (which is between D and E, but nearer to E) I find the Mixture is a Green inclining towards Blue; but becauſe 3 is near the Middle between the Center and the Circumference, the Colour is pretty much broken. To make the ſame thing more clear by another Example; ſuppoſe I would know what would reſult from a Mixture of two

Parts

Parts Yellow at P, three Parts Blue at Q, and five Parts Red at R. First I find the Place 3 of the Mixture of the Yellow and the Blue, as before. Then drawing the Line 3 R (because there are five Parts of the Colour at 3, and five Parts of the Colour at R) I divide it into ten Parts, and take the Point *r* five Parts distant from R. By this means *r* is the Center of Gravity of the three Colours at P, Q, and R, and is consequently the Place of the Mixture; which by drawing O*r* cutting the Circumference in *s*, I find to be an Orange a little inclining towards Red, and because *r* is much nearer the Center than the Circumference, the Colour is very much broken. And thus one may proceed in other Cases.

Again, having given the Place of any compound Colour, one may find what Colours may be mix'd to compound it. Thus having given the Colour at 3, drawing any Line P 3 Q thro' 3, the Colour proposed may be made by a Mixture of the Colours in P and Q, taking such a Proportion of them as is expressed by the Lines 3 P and 3 Q, that is, taking of the Colour P as much as in proportion to 3 Q, and as much of the Colour Q as is in proportion to 3 P. Or having drawn O 3 passing thro' the Points 1, 2, 4, the same Colour may be produced by mixing the Colours in 2 and 4 in proportion to the Lines 4.3 and 2.3; or it may be produced by breaking the simple Colour at 1 with White (which is at O) in the proportion of the Lines 3.1 and 3 O. And thus in other Cases.

4 The

The Proportions hitherto mention'd of the Colours to be uſed in the Mixtures, relate to the Quantity of the Rays of Light, and not to the Materials which artificial Colours are made of. Wherefore if ſeveral artificial Colours were to be mix'd according to theſe Rules, and ſome of them are darker than others, there muſt be a greater Proportion uſed of the darker Materials, to produce the Hue propoſed, becauſe they reflect fewer Rays of Light in proportion to their Quantities; and a leſſer Proportion muſt be uſed of the lighter Materials, becauſe they reflect a greater Quantity of Light.

If the Nature of the material Colours, which are uſed in Painting, was ſo perfectly known, as that one could tell exactly what Species of Colour, how perfect, and what degree of Light and Shade each Material has with reſpect to its Quantity, by theſe Rules one might exactly produce any Colour propoſed, by mixing the ſeveral Materials in their juſt Proportions. But tho' theſe Particulars cannot be known to ſufficient Exactneſs for this purpoſe, beſides the Tediouſneſs that would be in Practice, to meaſure the Colours according to their exact Proportions; yet the Knowledge of this Theory may be of great uſe in Painting. Suppoſe, for example, I had a Palate provided with the ſeveral Colours at *a*, *b*, *c*, *d*, *e*. Suppoſe for inſtance at *a* Carmine; at *b* Orpiment; at *c* Pink; at *d* Ultramarine, at *e* Smalts; and I had occaſion to make a broken Green, ſuch as I judge ſhould be placed at *x*. Looking round the Point *x*, I ſee that it does not lie a great

L deal

deal out of a Line drawn thro' *c* and *d*; therefore I conclude, that mixing the Colours *c* and *d* will come very near what I want. But becaufe *x* is nearer to the Center O than the Line *c d*, having brought my Tint as near as I can to what I want, fuppofe to *z*, I look from *z* crofs *x* for fome Colour oppofite to *z*, to break the Tint with, and I find the neareft to be *a*, therefore by mixing of the Colour *a* I bring the Compofition to the Tint I have occafion for. If the Colour *a* carries the Tint too much towards the Line OD, I put a little more of the Colour *d*, which brings it into the right place. Or having got the Tint *z*, I might have broken it with White, whofe place is at the Center O. Or putting a greater proportion of the Colour *d* inftead of *a*, I may afterwards break the Tint by means of the Colour *b*. And in the fame manner by only infpecting this Scheme, one may fee in what manner to make any Tints whatfoever, that can be produced by the Colours that one ufes. Thus one fees that Red and Yellow makes a broken Orange Colour, which may ftill be more broken by adding Blue, or Indico, or Violet, which are to be taken one or other, as one would have the Tint inclined more to the Yellow or to the Red; Blue bringing it toward the Yellow, and breaking it much; and Violet carrying it towards the Red, and not breaking it fo much.

From thefe Principles one may fee the Reafon why the Materials of the brighteft and fimpleft Colours are the moft valuable, and of them why the lighteft are moft to be efteem'd. The fimpleft

Colours

Colours are the moſt valuable, becauſe they cannot be produced by Mixture ; for Mixture always breaks the Colours. Suppoſe *a*, *b*, *c*, *d*, *e*, to be all the Colours you have, then drawing Lines to join the Points *a*, *b*, *c*, *d*, *e*, all the Tints that can be produced by thoſe Colours will have their Places within the Area of the Polygone *abcde*. That the lighter Colours are more valuable than the dark ones, is becauſe Black does not break the Colours ſo much as White ; ſo that it is eaſier to make the clean dark Tints with light Colours and Black, than to make the bright light ones, with dark Colours and White. For by what has been ſhew'd, White breaks the Colours very much, but Black being nothing but the abſence of Light, only darkens the Colours. Tho' upon account of the Imperfection of the Materials that are in uſe, Black does alſo break the Colours ſomething, becauſe there is no Material ſo perfectly black as to have no Colour at all, as one may ſee by the beſt Blacks having Lights and Shades. There will be other Exceptions alſo to be made in the application of theſe Obſervations to Practice, upon account of the particular Qualities of the Materials ſome Colours are made of. If all the Colours were as dry Powders, which have no effect upon one another, when mix'd, theſe Obſervations would exactly take place in the mixing of them. But ſome Colours are of ſuch a Nature, that they produce a very different effect upon their Mixture, to what one would expect from theſe Principles. So that it is poſſible there may be ſome dark Materials, which

when

when diluted with White, may produce cleaner and leſs compounded Colours than they gave when ſingle; as ſome Colours do very well to glaze with, which don't look well laid on in a Body. But theſe Properties of particular Materials I leave to be conſider'd by the Practitioners in this Art.

*

Fig. 2.

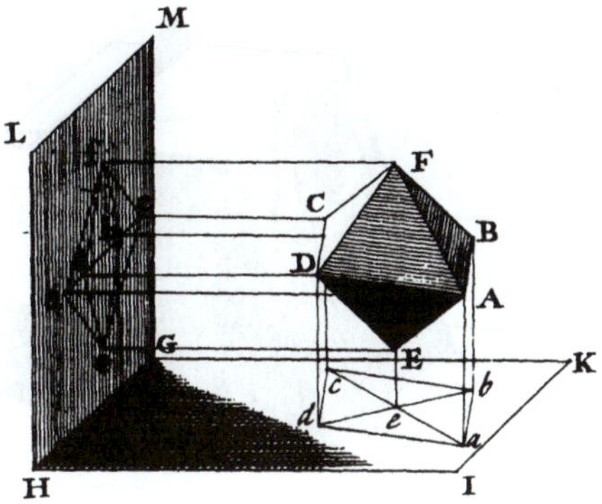

Fig. 1.

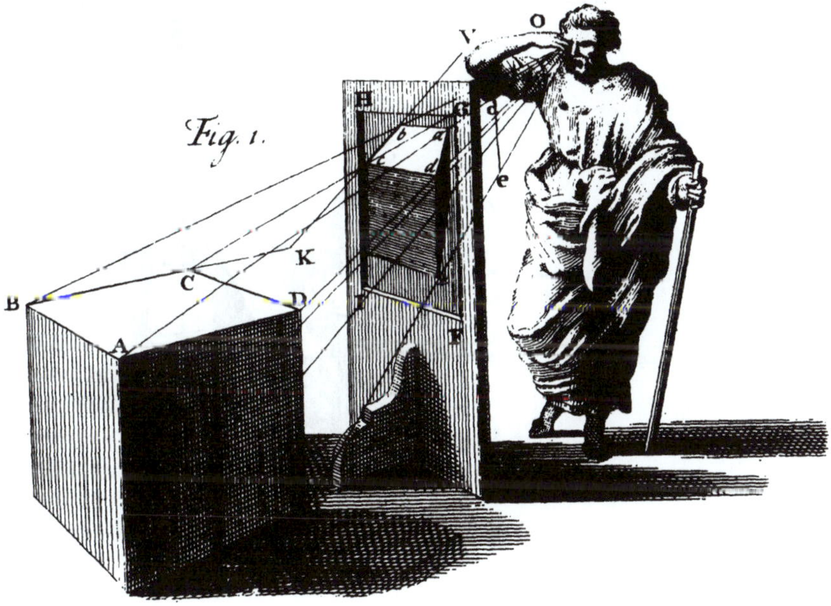

*

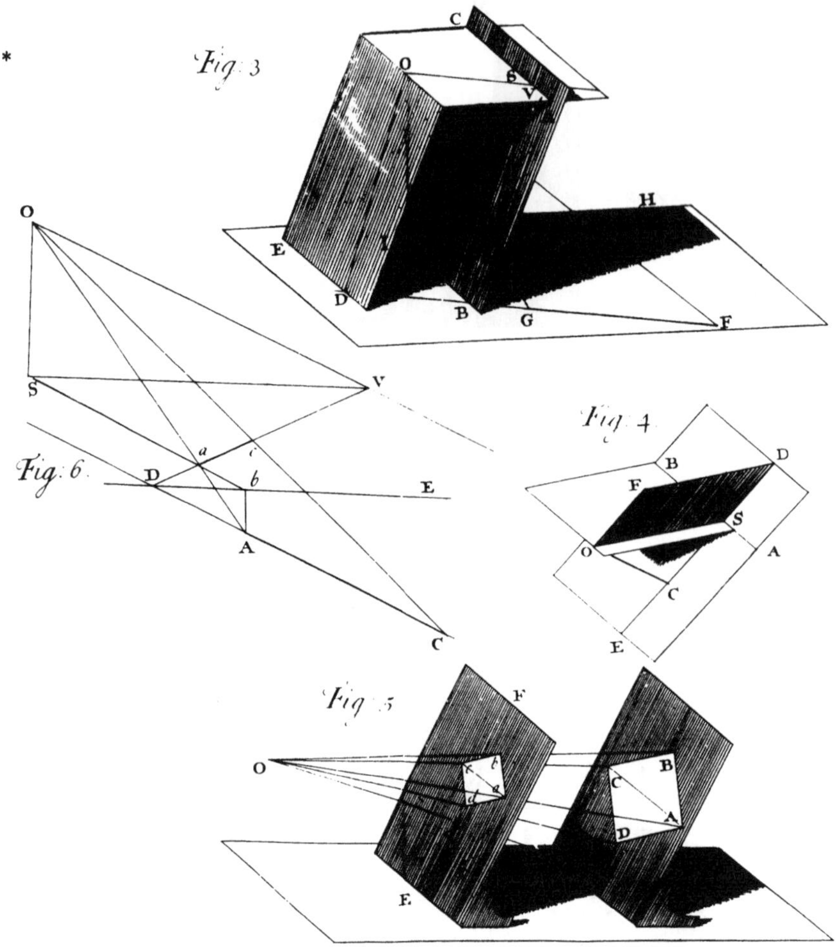
Fig: 3
C
O
S
V
H
E
D
B
G
F
O
S
V
Fig: 6
a
c
D
b
E
A
C
Fig: 4
B
D
F
S
O
A
C
E
Fig: 5
F
O
B
C
A
D
E

Fig: 8.

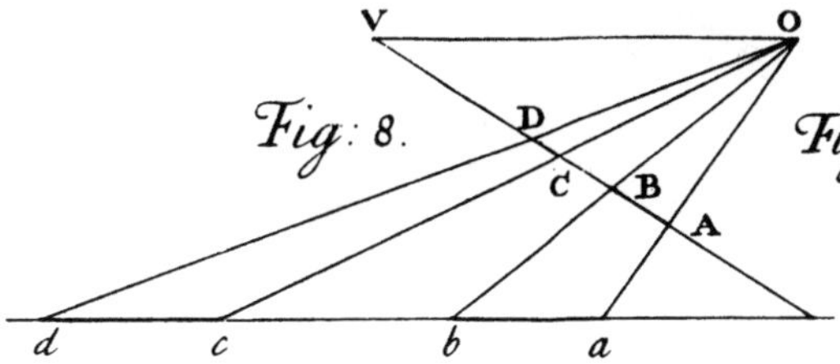

Fig: 7.

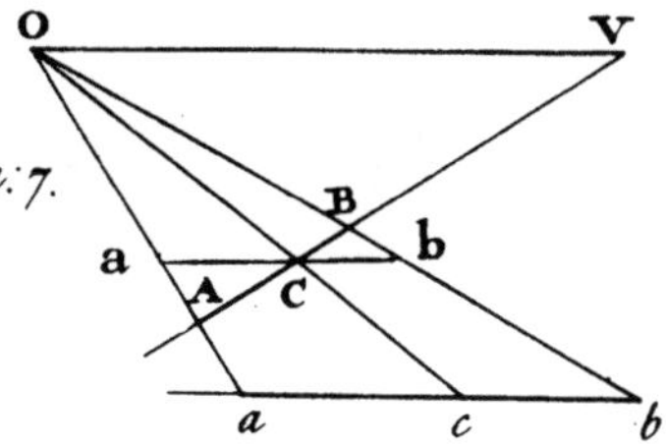

Fig: 9.

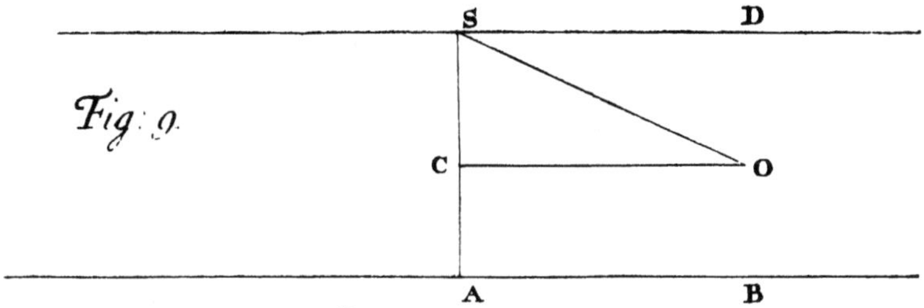

Fig: 10.

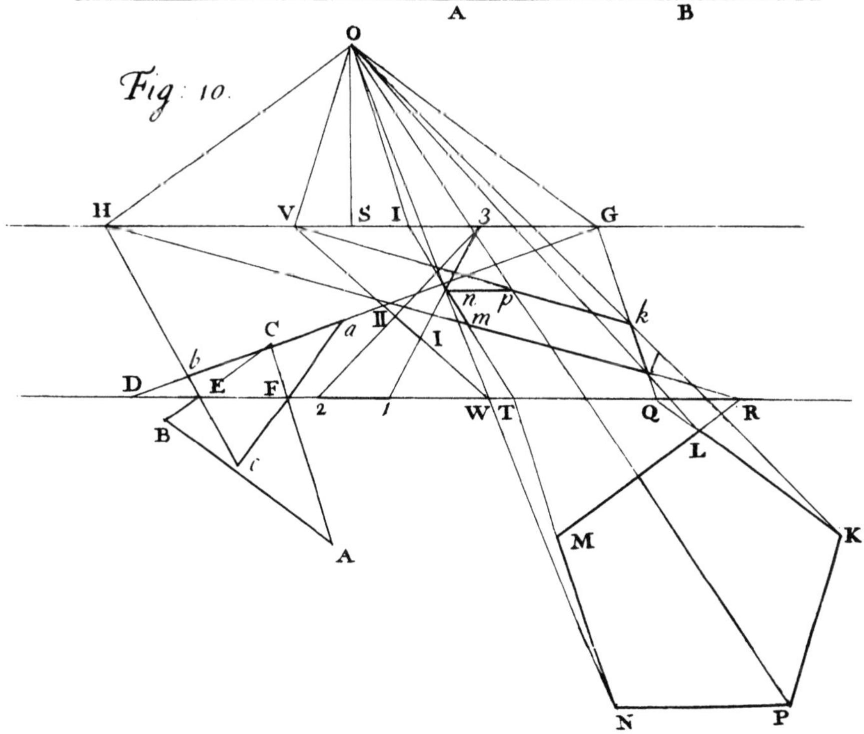

Fig: 11.

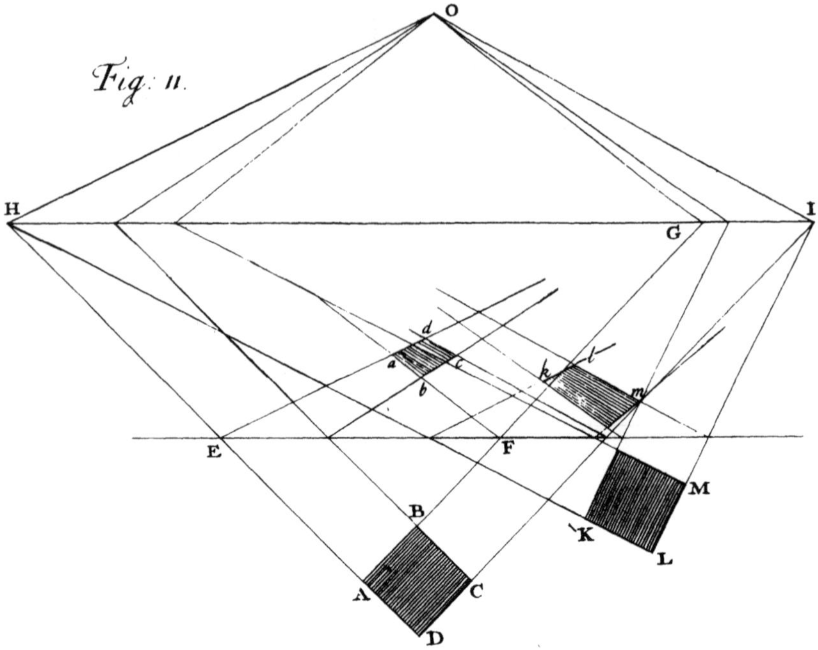

Fig: 12.

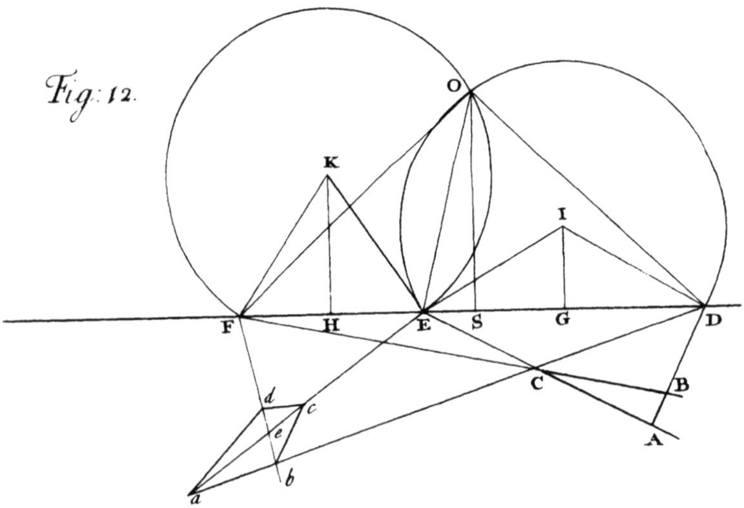

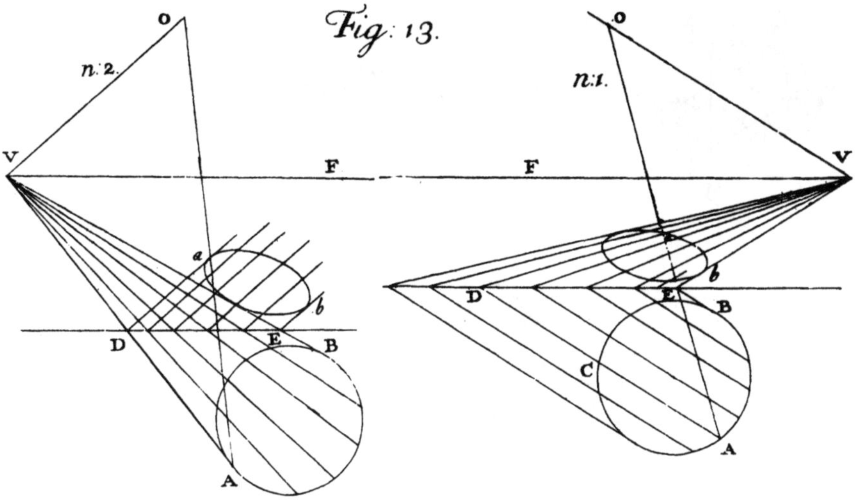

Fig: 13.

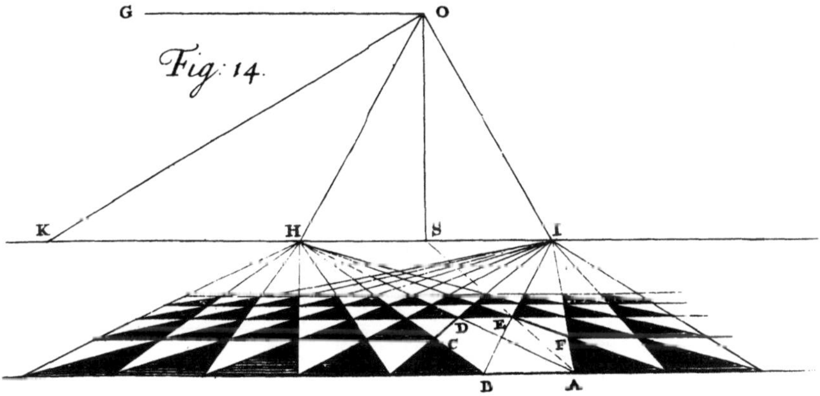

Fig: 14.

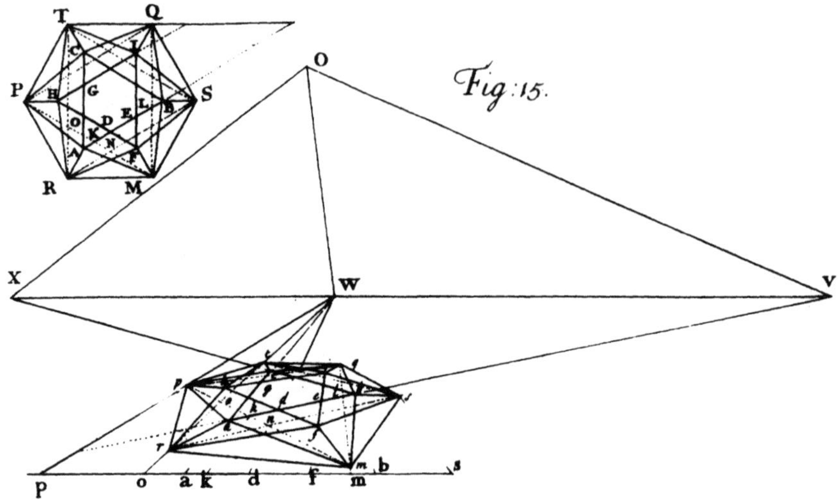

Fig: 15.

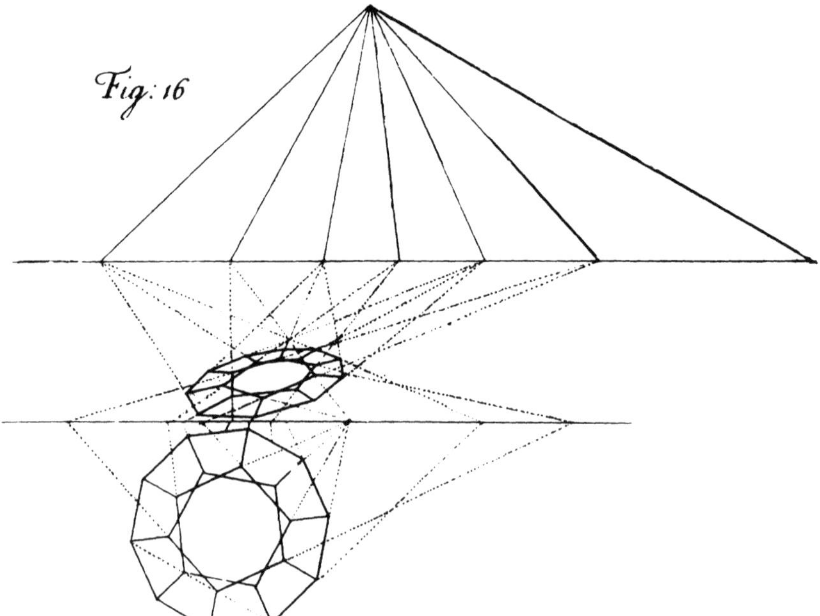

Fig: 16

Fig: 17.

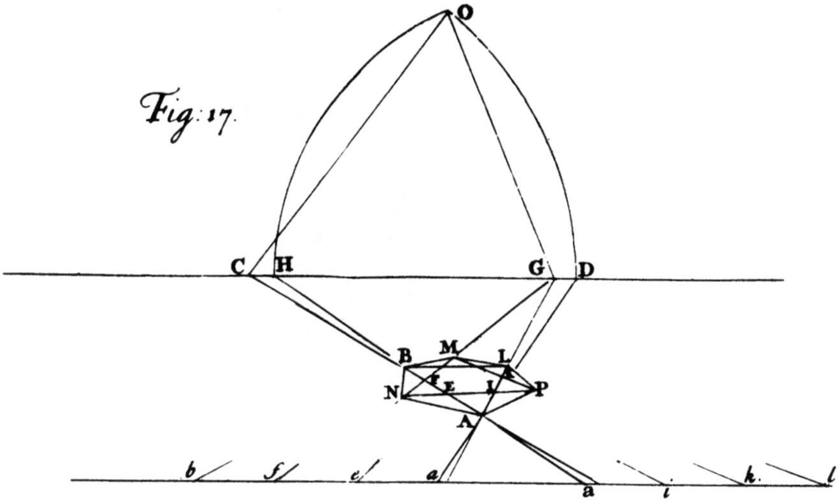

Fig: 18.

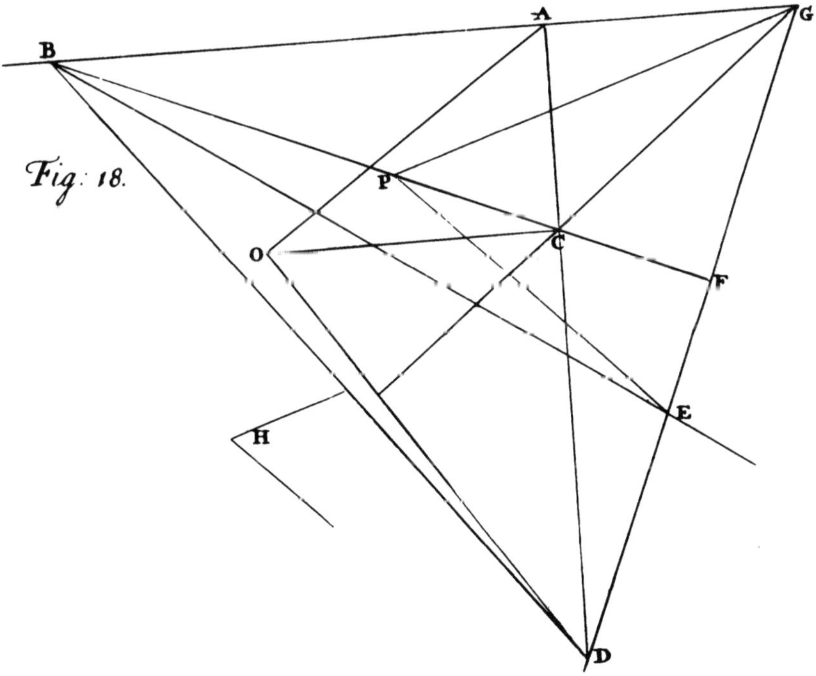

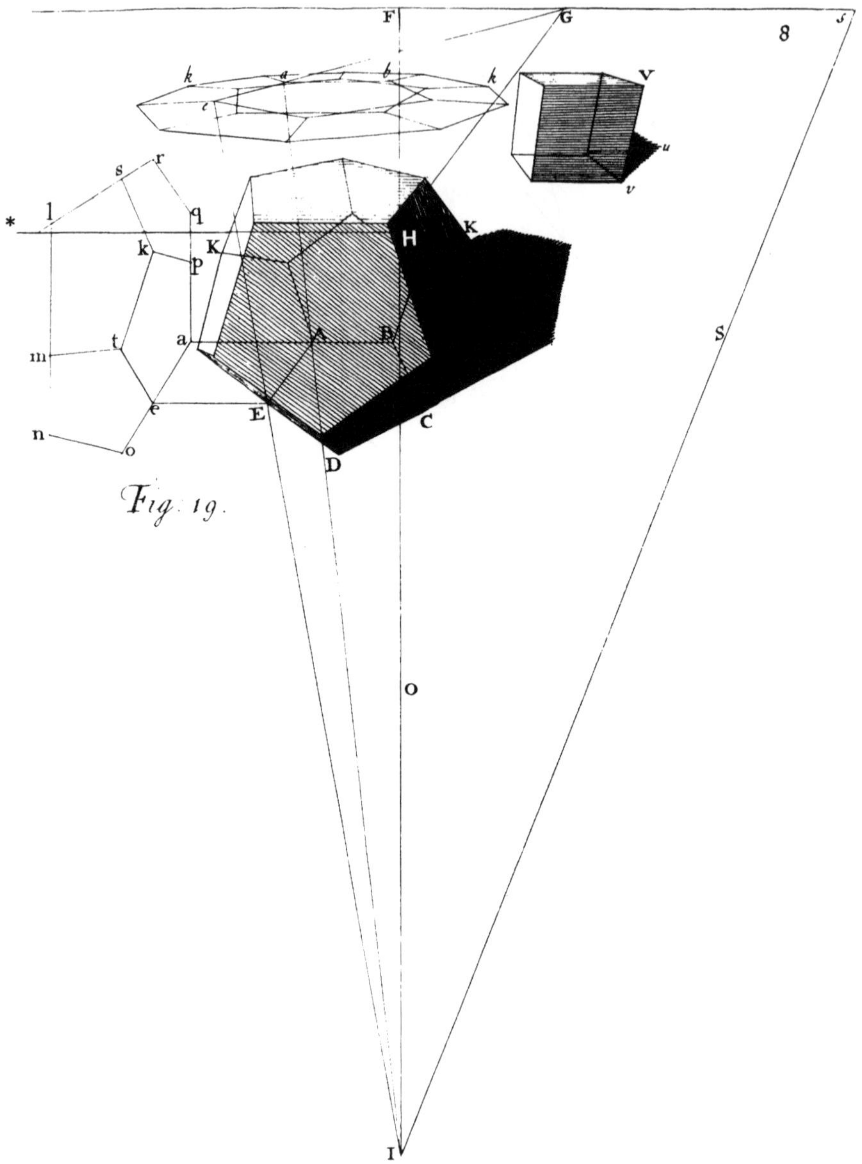

Fig. 19.

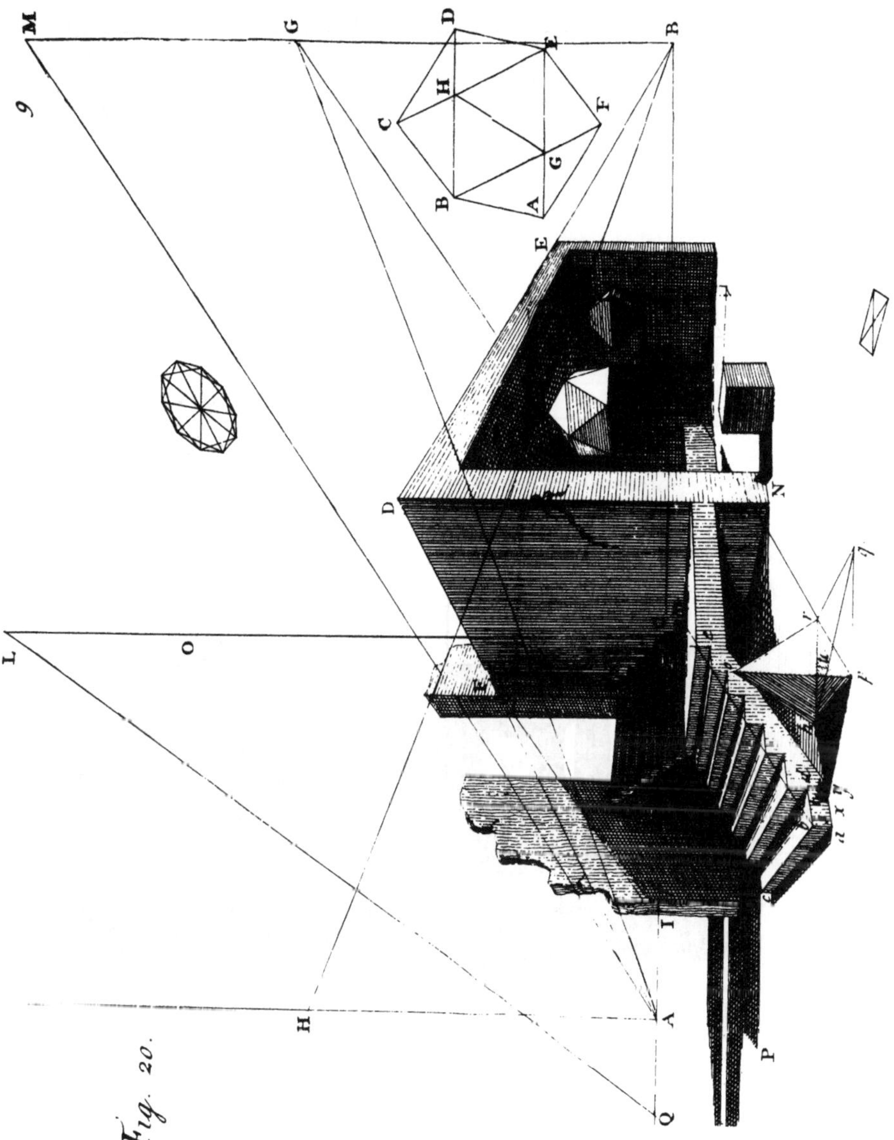

Fig. 20.

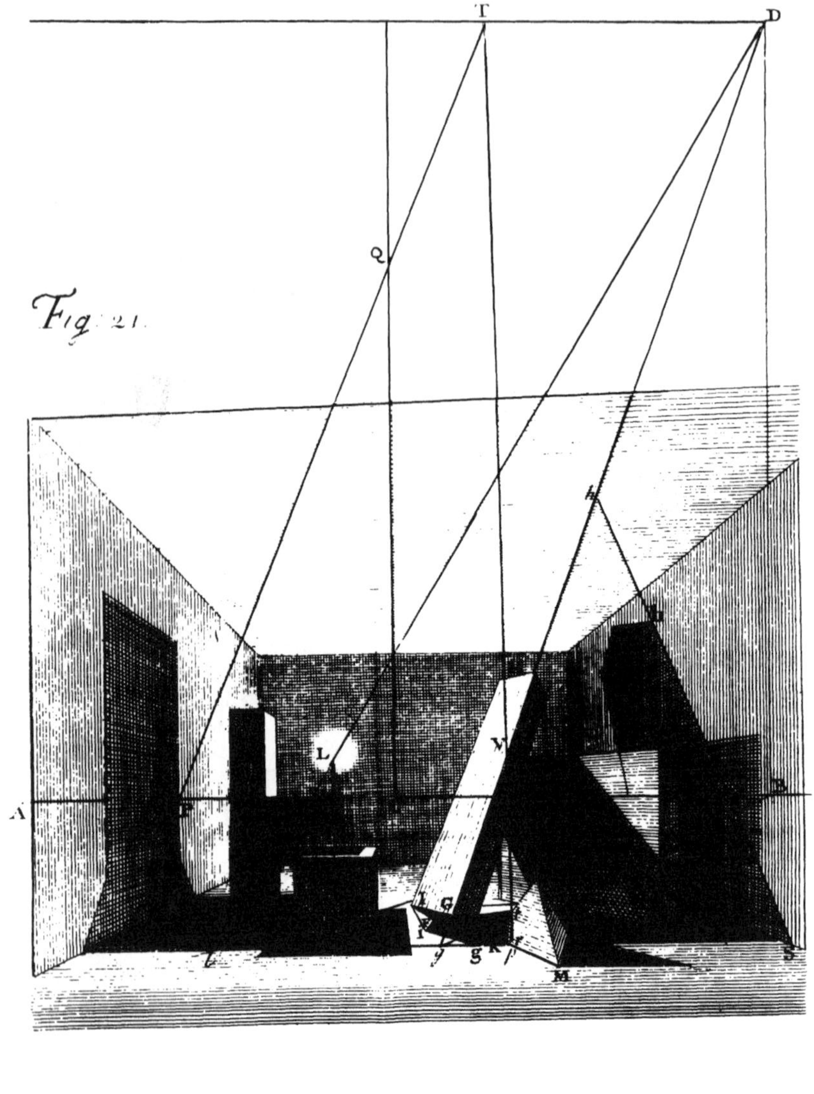

Fig: 21.

Fig: 22.

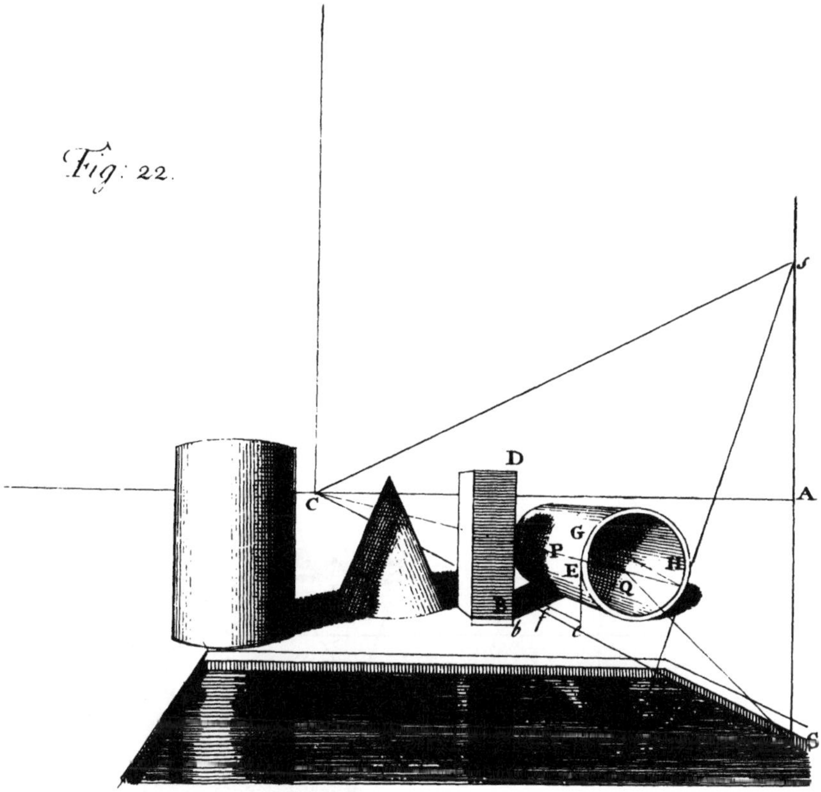

Fig: 23

13

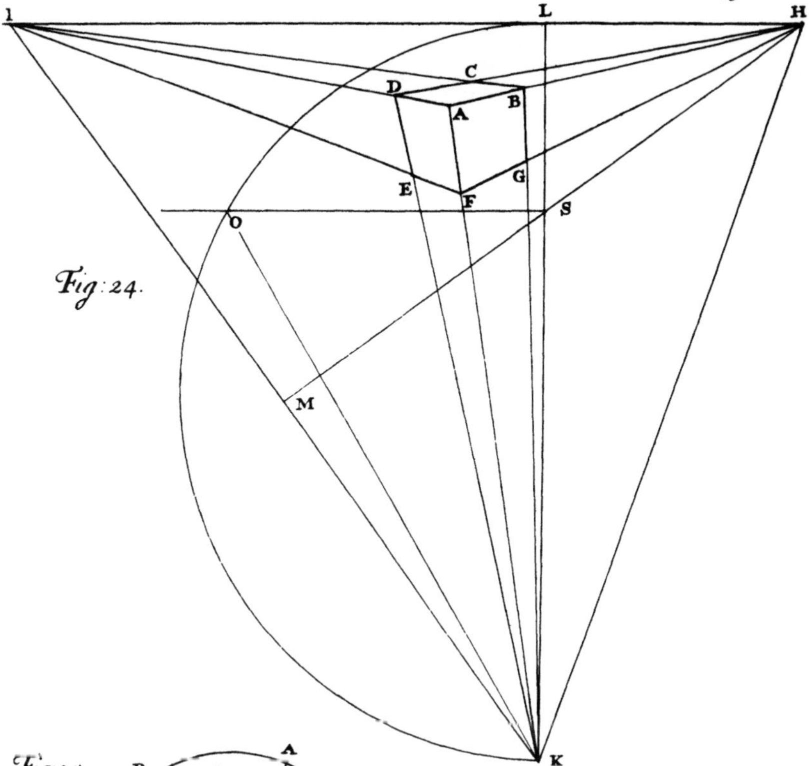

Fig: 24.

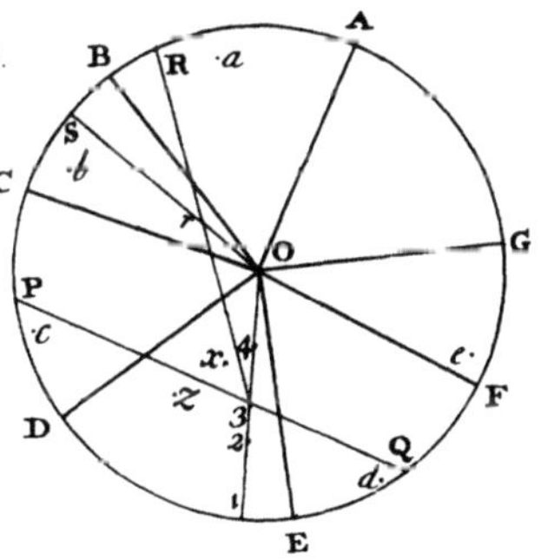

Fig: 25.

Notes to *New Principles*

(by Kirsti Andersen)

The references are to the page numbers of this book, the original page numbers are indicated in parentheses. A minus after a line number signifies that the counting is made from the bottom of the page.

1. p. 172 (12), l. 14: For Ichnographic read Schenographic.
2. p. 177 (17), l. 5^-: Since a directing point is only defined for lines that are not parallel to the picture plane (Definition XIII, p. 167) Taylor's definition of the *director* of a line, as the line joining the eye point and the directing point of the line, only applies to lines that are not parallel to the picture. Had he instead defined the director of a given line as the line of intersection of the directing plane and the plane defined by the eye and the given line he could also have used the definition for lines parallel to the picture.
3. p. 180 (20): Problem I, cf. pp. 13–14 and Problem 1, p. 84.
4. p. 183 (23): Problem III, cf. pp. 26–30 and Problem 9, p. 92.
5. p. 184 (24): Problem IV, cf. pp. 26–30 and Problem 10, p. 93.
6. p. 185 (25): Problem V, cf. Problem 3, p. 86.
7. p. 186 (26): Problem VI, cf. Example I, p. 96.
8. p. 187 (27), ll. 10 and 13: For *DF* and *AF* read *EF*.
9. p. 189 (29): Problem VII, cf. Example I, p. 96.
10. p. 189 (29), l. 8^-: Cf. the text to Figure 15, p. 22.
11. p. 191 (31), l. 4^-: The point **3** is the point which was later termed the measuring point for the line *I*, *II*. If we set $< VOS = v$, Taylor's construction is equivalent to determining the point **3** by $< SO\mathbf{3} = 45° - v/2$.
12. p. 192 (32): Problem XI, cf. pp. 26–27, 30 and Problem 8, p. 91.
13. p. 195 (35), ll. 18–30: Cf. Figure 29, p. 126, and the description p. 102.
14. p. 195 (35), l. 6^-: A line segment *AB* is cut in extreme and mean proportion (ratio) by the point *C* if

$$AB : AC = AC : CB.$$

15. p. 196 (36), ll. 3 and 5: For *e* and *l* read *f* and *m*.

16. p. 196 (36), l. 5^-: The construction of the plan of the dodecahedron is described p. 101, see also note 30, p. 138.
17. p. 197 (37), l. 4: For Projection read Projection of the Ichnography. The plan of an octahedron is also described on page 101.
18. p. 197 (37): Problem XIV, cf. pp. 34–36 and Problem 4, p. 87.
19. p. 199 (39): Problem XV, cf. pp. 34, 37 and Problem 5, p. 88.
20. p. 200 (40): Problem XVI, cf. Problem 6, p. 89.
21. p. 201 (41): Problem XVII, cf. pp. 37–38 and Problem 7, p. 90.
22. p. 202 (42), l. 14^-: For *DPB* read *EPB*; l. 12^-: for *DB* read *EB*; l. 7^-: for *BD* read *BE*.
23. p. 204 (44): Example V, cf. pp. 39–41; *Linear Perspective* (p. 101 and Figure 28, p. 126) contains a description of the construction of the plan and elevation of the dodecahedron. The elevation corresponding to the perspective image in Figure 19 (p. 238) is illustrated in Figure 1.
24. p. 206 (46), ll. 7–14: In this section Taylor mentioned other properties of the elevation of the dodecahedron (Figure 1) than he did in *Linear Perspective*, cf. the previous note. The former can be derived from the latter; in particular, the result concerning extreme and mean proportion (ratio) is a consequence of relations (1) and (2) in Note 30, p. 139.
25. p. 206 (46), l. 1^-: The base *hrp* is constructed, as we are informed, p. 207, l. 12, by Problem 12.
26. p. 207 (47), l. 1: The point *I* is not defined, but seems to be the vanishing point of *hp*. l. 4: for *kpr* read *hpr*; l. 9: for *VX* read *KL*; l. 10: for *upr* read *opr*; ll. 17–21: cf. Example II, p. 195.
27. p. 211 (51): Example IX, cf. pp. 42–45.
28. p. 215 (55), l. 8: For Figure 17 read Figure 7.
29. p. 217 (57): Problem XXII, cf. p. 46.

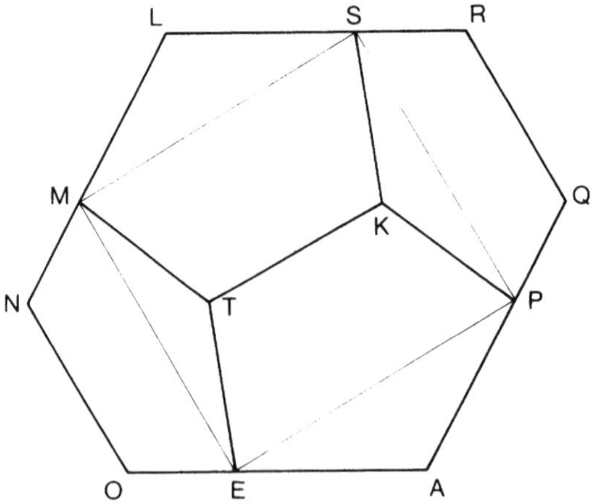

Figure 1. The elevation of the dodecahedron which is thrown in perspective in Figure 19 (p. 238).

30. p. 217 (57): Problem XXIII, cf. pp. 46–47.
31. p. 219 (59): Appendix I, cf. p. 48.
32. p. 222 (62): Appendix II, cf. pp. 48–51.
33. pp. 231–243: The format of this book has made it necessary to reduce the figures from *New Principles*, and to reduce them in different scales.
34. p. 232: Figure 3 contains some mistakes, for a correction, see page 11.
35. p. 238: A white H has been added to the figure to make the point *H* visible.

Bibliography

Aguilon, François

1613: *Opticorum libri sex*, Antwerp.

Alberti, Leon Battista

1435: *De pictura.*

1436: *Della pittura.*

The first printed text was a Latin edition appearing in Basel 1540. Later many editions and translations have been published. Among the recent ones is:

1972: *On Painting and On sculptures*, ed. Cecil Grayson, London.

Aleaume, Jacques

1643: *La perspective speculative et pratique ... sans plan géométral & sans tiers poinct*, ed. Étienne Migon, Paris.

This book has a complicated printing history which is explained on pages 156–157. The royal engineer Aleaume died in 1627 and left a manuscript on perspective. This was bought by the printer Charles Hulpeau who wanted to publish it; he made woodcuts of the figures and typeset part of the manuscript before death stopped his work. In 1641 Migon took upon himself the job of having Hulpeau's work finished, and he produced *La perspective speculative* .... However, in the process of editing the book he decided to change many things, including the figures—without indicating his changes and additions. Thus it is completely unclear who is the author of which part of the book.

Andersen, Kirsti

1984: "Some Observations Concerning Mathematicians' Treatment of Perspective Constructions in the 17th and 18th Centuries," *Mathemata, Festschrift für Helmuth Gericke*, eds. M. Folkerts & U. Lindgren, Stuttgart, pp. 409–425.

1987_1: "Ancient Roots of Linear Perspective", *From Ancient Omens to Sta-*

tistical Mechanics. Essays on the Exact Sciences Presented to Asger Aaboe, eds. J.L. Berggren & B.R. Goldstein, Copenhagen, pp. 75–89.

1987_2: "The Problem of Scaling and of Choosing Parameters in Perspective Constructions, Particularly in the One by Alberti," *Analecta Romana* **16**, pp. 107–128.

Auchter, Heinrich

1937: *Brook Taylor der Mathematiker und Philosoph. Beiträge zur Wissenschaftsgeschichte der Zeit des Newton–Leibniz Streites*, Würzburg.

Bardwell, Thomas

1756: *The Practice of Painting and Perspective Made Easy*, London.

Bkouche, Rudolf

1990: *La naissance du projectif. De la perspective à la géométrie projective*, IREM de Lille.

Blacker, George O.

1885–1888: *John Heywood's Second Grade Perspective ... adapted from Dr. Brook Taylor*, Manchester.

Bosse, Abraham

1643_1: *La maniere universelle de M. Desargues ... pour poser l'essieu et placer les heures ... aux cadrans du soleil*, Paris.

1643_2: *La pratique du trait à preuves de M. Desargues ... pour la coupe des pierres ...*, Paris.

1653: *Moyen universel de pratiquer la perspective sur les tableaux, ou surfaces irregulieres ...*, Paris.

1667: *La peinture convertie aux precises regles de son art ...*, Paris.

Bourgoing, Charles

1661: *La perspective affranchie, contenant la vraye et naturele pratique ... sans tracer ny supposer le plan geometral ordinaire*, Paris.

Bricard, Raoul

1924: *Petit traité de perspective*, Paris.

Catalogue

1732: *A Catalogue of the Libraries of Joseph Hall Esq. late one of the Six Clerks in Chancery and of Brook Taylor, LL.D and F.R.S. lately deceased ... which will be sold tuesday the 22^d day of February 1731–2, by Fletcher Gyles*, London.

Chasles, Michel

1837: *Aperçu historique sur l'origine et le développement des méthodes en géométrie*, Bruxelles. Second edition, Paris 1875.

Coolidge, Julian L.

1963: *A History of Geometrical Methods*, New York. First edition, Oxford 1940.

Cowley, John Lodge

1766: *The Theory of Perspective*, London.

Cremona, Luigi

1865: (using as a pseudonym the anagram Marco Uglieni of his name) "I principii della prospettiva lineare secondo Taylor," *Giornale di matematiche* **3**; also as a separate publication Bologna 1865.

Dechales, Claude François Milliet

1674: *Cursus seu mundus mathematicus*, 3 vols., Lyon. Second edition by Amati Varcin, 4 vols., Lyon 1690.

De Morgan, Augustus

1861: "Notes on the History of Perspective," *The Athenaeum* (Nov. 30), pp. 727–728.

Desargues, Girard

1636: *Exemple de l'une des manieres universelles du S.G.D.L. touchant la pratique de la perspective sans emploier aucun tiers point ...*, Paris. Facsimile and an English translation in Field and Gray 1987.

Diderot, Denis et D'Alembert, Jean le Rond (eds.)

1780: *Encyclopédie, ou dictionnaire raisonné des sciences, des arts et des métiers*, tome XXV, Lausanne and Berne.

Ditton, Humphry

1712: *A Treatise of Perspective, Demonstrative and Practical*, London.

Dubreuil, Jean

1642: *La perspective pratique ... Par un religieux de la compagnie de Jesus*, Paris. Published anonymously.
Several later French editions, translation into German and two into English, the first being:

1672: *Perspective practical*, ed. Robert Pricke, London.

Edwards, Edward

1803: *A Practical Treatise of Perspective, on the Principles of Dr. B. Taylor*, London.

Emerson, William

1765: *Cyclomathesis: or an Easy Introduction to the several Branches of the Mathematics*, vol. VI, London.

Euclid

1908: *The Thirteen Books of Euclid's Elements*, ed. T.L. Heath, Cambridge. Several later editions.

Feigenbaum, Lenore

1985: "Brook Taylor and the Method of Increments," *Archive for History of Exact Sciences* **34**, pp. 1–140.

1986: "Happy Tercentenary, Brook Taylor!," *The Mathematical Intelligencer* **8**, no. 1, pp. 53–56.

Ferguson, James

1775: *Art of Drawing in Perspective Made Easy ...*, London.

Field, J.V. and Gray, Jeremy J.

1987: *The Geometrical Work of Girard Desargues*, New York, etc.

Fournier, Daniel

1761: *A Treatise on the Theory and Practice of Perspective. Wherein the Principles ... by Dr. B. Taylor are explained by the means of moveable Schemes*, London. Later editions, London 1762, 1763, 1764.

's Gravesande, Willem Jacob

1711: *Essai de perspective*, La Haye. Reprinted in 's Gravesande 1774. English translation:

1724: *An Essay on Perspective ...*, ed. E. Stone, London.

1774: *Oeuvres philosophiques et mathématiques de Mr G.J. 's Gravesande*, vol. I, Amsterdam.

Guidobaldo del Monte

1600: *Perspectivae libri sex*, Pesaro. Italian translation in:

1984: *I sei libri della prospettiva di Guidobaldo dei marchesi Del Monte*, ed. R. Sinisgalli, Roma.

Halfpenny, Willam

1731: *Perspective Made Easy*, London.

Hamilton, John

1738: *Stereography, or a Compleat Body of Perspective*, London. Later editions, London 1740, 1748, 1749.

Highmore, Joseph

1754: *A Critical Examination of those two Paintings on the Ceiling of the Banqueting-House at Whitehall, in which Architecture is introduced, so far as relates to the Perspective*, London.

1763: *The Practice of Perspective. On the Principles of Dr. Brook Taylor, written many years since, but now first published*, London

Huret, Grégoire

1670: *Optique de portraiture et peinture*, Paris.

Jones, Phillip S.

1947: *The Development of the Mathematical Theory of Linear Perspective and its Connections with Projective and Descriptive Geometry with Especial Emphasis on the Contributions of Brook Taylor*. Dissertation. University of Michigan.

1950: "Brook Taylor and the Mathematical Theory of Linear Perspective, his Contributions and Influence," *Proceedings of the International Congress of Mathematicians*, vol. II.

1951: "Brook Taylor and the Mathematical Theory of Linear Perspective," *American Mathematical Monthly* **58**, pp. 597–606.

1976: "Taylor, Brook." *Dictionary of Scientific Biography* (ed. C.C. Gillispie), vol. 13, New York, pp. 265–268.

Kerber, Bernhard

1971: *Andrea Pozzo*, Berlin and New York.

Kirby, John Joshua

1754: *Dr. Brook Taylor's Method of Perspective Made Easy, Both in Theory* and Practice. Ipswich. Later editions, Ipswich 1755, London 1765, 1768.

s.a.: *Dr. Brook Taylor's Method of Perspective, compared with Examples lately publish'd on this Subject as Sirigatti's by Isaac Ware*, London. According to De Morgan, 1861 the book was published in 1757.

1761: *The Perspective of Architecture ... deduced from the Principles of Dr. Brook Taylor*, London.

La Caille, N. L.

1756: *Leçons élémentaires d'optique*, second edition, Paris.

Lambert, Johann Heinrich

1759_1: *Die freye Perspektive*, Zürich. Second edition part of Lambert 1774. Reprinted in Lambert 1943. French edition:

1759_2: *La perspective affranchie de l'embaras du plan géométral.* Zürich. Facsimile, Paris 1977.

1774: *Die freye Perspective ... mit Anmerkungen und Zusätzen vermehrt*, Zürich. Reprinted in Lambert 1943. French translation of the *Anmerkungen* in Laurent 1987.

1916: "Johann Heinrich Lamberts Monatsbuch," ed. K. Bopp, *Abhandlungen der Königlichen Bayerischen Akademie der Wissenschaften. Mathematisch–physikalische Klasse*, XXVII, Band 6.

1943: *Johann Heinrich Lambert, Schriften zur Perspektive*, ed. Max Steck, Berlin.

1981: *Essai sur la perspective*, translation of *Anlage zur Perspektive* (1752), ed. Roger Laurent and Jeanne Peiffer, Coubron.

Lamy, Bernard

1701: *Traité de perspective*, Paris. English translation.

1710: *Perspective Made Easie*, ed. A. Forbes, London.

Laurent, Roger

1987: *La place du J.-H. Lambert (1728–1777) dans l'histoire de la perspective. Suivi de la version intégrale ... des Notes et additions à la perspective affranchie ...* . Traduction de Jeanne Peiffer, Paris.

Leonardo da Vinci

1651: *Trattato della pittura*, ed. R. du Fresne, Paris.

1721: *A Treatise of Painting*, London.

Loria, Gino

1908: "Perspektive und darstellende Geometrie" in Moritz Cantor, *Vorlesungen über Geschichte der Mathematik*, vol. 4, Leipzig, pp. 577–637.

Maignan, Emmanuel

1648: *Perspectiva horaria sive horographiae gnomonicae*, Roma.

Malton, James

1800: *The Young Painter's Maulstick: being a Practical Treatise on Perspective founded on the process of Vignola and Sirigatti, ... united with the theoretic principles of ... B. Taylor*, London.

Malton, Thomas
1775: *A Compleat Treatise on Perspective in Theory and Practice on the True Principles of Dr. Brook Taylor, Made Clear ...*, London. Also London 1776, 1778 and 1779.
1783: *An Appendix or Second Part to the Compleat Treatise on Perspective* containing a brief *History of Perspective*, London. Second edition, London 1800.

Marolois, Samuel
1614: *Opera mathematica/Oeuvres mathematiques traitans de géométrie, perspective ...*, Den Haag.
This work was reprinted often, among the years of publication are 1628, 1638, 1647, and 1662; it appeared in Dutch, German, French, and Latin. There is some language confusion in several of the editions, their title pages being in one language and the main text in another.

Michel, S. N.
1771: *Traité de perspective linéaire*, Paris.

Montucla, Jean Étienne
1758: *Histoire des mathématiques*, 2 vols. Paris. Second edition, 4 vols., Paris 1799–1802, reprinted, Paris 1968.

Moxon, Joseph
1670: *Practical Perspective; or Perspective Made Easie*, London.

Newton, Isaac
1704: *Opticks: Or, a Treatise of the Reflexions, Refractions, Inflexions and Colours of Light*, London. Facsimile, Bruxelles 1966.

Noble, Edward
1771: *The Elements of Linear Perspective*, London.

Ozanam, Jacques
1693: *Cours de mathématique*, 5 vols., Paris. Later editions, Amsterdam 1697, 1699. Translated as *Cursus mathematicus: or a Compleat Course of the Mathematicks*, 5 vols., London 1712.

Poudra, N. G.
1864: *Histoire de la perspective ancienne et moderne*, Paris.

Pozzo, Andrea
1693–1700: *Perspectiva pictorum et architectorum*, 2 parts, *pars prima* Roma 1693, *pars secunda* Roma 1700, in Latin and Italian.
Through the eighteenth century these volumes were reissued very frequently either separately or together. Moreover, they were translated into many languages—including Chinese and modern Greek. For an extensive bibliography, see Kerber 1971, pp. 267–270. Like the original the translations often had the text in two languages. There exist at least six English editions, the first one of the first volume being:
1707: *Rules and Examples of Perspective*, in Latin and English, ed. John James, London.

Priestley, Joseph
1770: *A familiar Introduction to the Theory and Practice of Perspective*, London.

Scheiner, Christoph
1631: *Pantographice*, Roma.

Schooten, Frans van
1660: "Tractaet der perspective ...," published in *Mathematische Oeffeningen*, Amsterdam.

Schüling, Hermann
1973: *Theorien der malerischen Linear-Perspektive vor 1601*, Giessen.

Serlio, Sebastian
1584: *Tutte l'opere d'architettura e prospettiva ...* Venezia.
Several editions in Italian and other languages; English translation:
1611: *The Second Booke of Architecture Made by S. Serlij*, London.

Shapiro, Alan E.
forthcoming:
Newton on the Nature and Rules "of Color Mixing."

Sinisgalli, Rocco
1978: *Per la storia della prospettiva (1405–1605). Il contributo di Simon Stevin allo sviluppo scientifico della prospettiva artificiale ed i suoi precedenti storici*, Roma.

Sirigatti, Lorenzo
1596: *La pratica di prospettiva*, Venezia. English translation:
1756: *Practice of Perspective with Figures Engraved by Isaac Ware*, London.

Stellini, Giacopo
1782: *Opere varie*, vol. III (contenente alcuni opuscoli mathematici), ed. Antonio Evangel, Padova.

Stevin, Simon
1605_1: "Van de verschaeuwing. Eerste bouck der deursichtighe" in *Derde Stuck der Wisconstighe Ghedachtnissen. Van de Deursichtighe*," Leiden.
Translations in:
1605_2: *Hypomnemata mathematica a Simone Stevino*. Tomus tertius, ed. W. Snellius, Leiden.
1605_3: *Mémoires mathématiques par Simon Stevin*. Livre trois, ed. J. Tuning, Leiden.
Slightly revised version of this translation in:
1634: *Les oeuvres mathématiques de Simon Stevin, où sont inserées les mémoires mathématiques ...*, ed. A. Girard, Leiden.
The Dutch text has been reprinted together with an English translation in:
1958: *The Principal Works of Simon Stevin*, ed. D.J. Struik, vol. II.B, Amsterdam.
Furthermore, the Latin text has been republished together with an Italian translation in Sinisgalli 1978.

Tacquet, Andreas

1669: *Opera mathematica*, Antwerp. Second edition 1707.

Taylor, Brook

1715_1: *Linear Perspective: or, a New Method of Representing justly all Manner of Objects*, London. Facsimile edition, New York, etc. 1991.

1715_2: "Accounts of Books: *Linear Perspective* ... By Brook Taylor ... London 1715," *Philosophical Transactions*, **29**, pp. 300–304.
Published anonymously but its contents clearly show that Taylor is the author.

1715_3: *Methodus incrementorum directa et inversa*, London. Second edition, London 1717.

1719: *New Principles of Linear Perspective: or the Art of Designing on a Plane the Representations of all sorts of Objects, in a more General and Simple Method than has been done before*, London.
Reissued as the third edition (because Taylor 1715 is counted as the first) by J. Colson 1749. Edited as *Dr. Brook Taylor's Method of Perspective* by I. Ware, London 1767. A revised edition, London 1811. *Dr. Brook Taylor's Principles of Linear Perspective*, ed. J. Jopling, London 1835. Facsimile of the first edition New York, etc. 1991.
Translations:

1755: *Elementi di perspettiva secondo li principii di Brook Taylor, con varieii aggiunti*, ed. François Jacquier, Roma.

1757: *Nouveaux principes de la perspective linéaire, traduction de deux ouvrages, l'un anglois de Docteur Brook Taylor, l'autre latin de M. Patrice Murdoch*, Amsterdam. Second edition, Amsterdam 1759.
The translator's name is not mentioned, but there is a general agreement that the translation was made by Antoine Rivoire.

1782: *Nuovi principii della prospettiva lineare*, published in Stellini 1782.

1793: *Contemplatio philosophica. A Posthumous Work of the Late Brook Taylor ... To which is Prefixed a Life of the Author, by His Grandson Sir William Young ...*, London.

Vagnetti, Luigi

1979: *Prospettiva. De naturali et artificiali perspectiva—bibliografia ragionata delle fonti teoriche e delle ricerche di storia della prospettiva; contributo alla formazione della conoscenza di un'idea razionale, nei suoi sviluppi da Euclide a Caspar Monge*, Firenze (Studi e documenti di architettura, n. 9–10).

Index

MIX
Papier aus verantwortungsvollen Quellen
Paper from responsible sources
FSC® C105338

If you have any concerns about our products, you can contact us on
ProductSafety@springernature.com

In case Publisher is established outside the EU, the EU authorized representative is:
Springer Nature Customer Service Center GmbH
Europaplatz 3, 69115 Heidelberg, Germany

Printed by Libri Plureos GmbH
in Hamburg, Germany